QUANTITATIVE STRUCTURE-ACTIVITY RELATIONSHIP

Edited by **Fatma Kandemirli**

Quantitative Structure-activity Relationship
http://dx.doi.org/10.5772/65597
Edited by Fatma Kandemirli

Contributors

Anand Gaurav, Ming Xing, Mayasah Al-Nema, Francisco Torrens, Gloria Castellano, Rina Herowati, Gunawan Pamudji Widodo, Yoshio Hamada, Yoshiaki Kiso, Fatma Kandemirli

Published by InTech

Janeza Trdine 9, 51000 Rijeka, Croatia

Publishing Process Manager Mirena Calmic
Technical Editor SPi Global
Cover InTech Design team

First published August, 2017
Printed in Croatia
Legal deposit, Croatia: National and University Library in Zagreb

Additional hard copies can be obtained from orders@intechopen.com

Quantitative Structure-activity Relationship, Edited by Fatma Kandemirli
p. cm.
Print ISBN 978-953-51-3409-1
Online ISBN 978-953-51-3410-7

For EU product safety concerns:
IN TECH d.o.o., Prolaz Marije Krucifikse Kozulić 3, 51000 Rijeka, Croatia,
info@intechopen.com or visit our website at intechopen.com.

Meet the editor

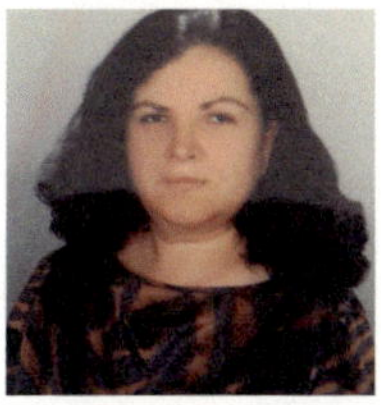

Fatma Kandemirli obtained her PhD at Gebze High Technology Institute, Chemistry Department in 1999. Her academic education and positions include İstanbul Technical University Chemical Engineering Department (1976); 2005–2010 Associate Professor, Department of Chemistry, University of Kocaeli, Kocaeli, Turkey; 2010–2012 Professor, Department of Chemistry, University of Niğde, Niğde, Turkey; 2012 Professor, Department of Chemistry, University of Kastamonu, Kastamonu, Turkey; 2012 Professor, Department of Biomedical Engineering, University of Kastamonu, Kastamonu.
Dr. Kandemirli's activities and interests are: Synthesis of inorganic compounds, QSAR, reaction mechanism, quantum chemical calculations, quantum chemical calculations of corrosion inhibitors.
She has published 95 peer-reviewed scientific papers. She has been cited without self-citations (ISI) 826 times.

Contents

Preface

The basic aim of a quantitative structure-activity relationship (QSAR) is to find a statistical model that is related to a structure of a chemical, which is described by reference to certain parameters and to its biological activity. The QSAR analysis allows quantitative analysis of the interactions between chemicals and life, and it has been successfully applied in many different areas including pharmaceutical, agricultural, molecular, and cellular events underlying the ozone toxicity in the lung, the prediction of toxicity to environmental species, the identifying of hazardous compounds for screening of inventories of present compounds, and the risk assessment and inhibition of corrosion.

QSAR methods can contribute to billion errors, boosting cost-effectiveness and time effectiveness of risk assessment procedure; saving of million test animals; elucidation of mechanisms; the identification of toxic chemicals in chemical structure, primarily, the prediction of specific properties of chemicals based on the structure of the substances; the design of safer chemicals; and decision-making process of deciding whether it is necessary to elucidate the concern of a test point.

In the second chapter, application of computational methods used in discovery and development of phosphodiesterase inhibitors will be presented.

In the third chapter, it will be designed that potent small-molecule non-peptidic BACE1 inhibitors with the use of hypothesized that the interaction of the P2 position of the inhibitor with the S2 site of BACE1 was critical for the mechanism of inhibition and with the propose the novel concept of "electron donor bioisostere" for drug discovery.

The chromatographic retention times of chlorogenic acids in coffee will be modeled by structure-property relationships in the forth chapter. An extension of solvent-dependent conformational analysis program (SCAP) octanol-water model to organic solvents was used in this study.

In the fifth chapter, to investigate the interaction modes of natural compounds to the potential macromolecular targets, docking simulation is going to be performed by using AutoDock Vina.

Fatma Kandemirli
Faculty of Engineering and Architecture
Biomedical Engineering Department
University of Kastamonu
Kastamonu, Turkey

Chapter 1

Introductory Chapter: Some Quantitative Structure Activity Relationship Descriptor

Fatma Kandemirli

Additional information is available at the end of the chapter

http://dx.doi.org/10.5772/intechopen.69642

1. Quantitative structure activity relationship

The Quantitative structure–activity relationship (QSAR) specifies the function between any property of the system under examination and the molecular system and its any geometric and chemical characteristics. QSAR tries to find a relationship between activity and molecular characterization so that these functions can be used to calculate the property of the new compounds.

QSAR models are available at the intersection of chemistry, statistics and property of the system. This property can be activity inhibition and so on. These requirements for the creation of the QSAR model are a data set, providing experimental measurements for the system. These datasets typically consist of hundred or fewer compounds associated with a specific parameter such as inhibition efficiency, intestinal absorption, volume of distribution, blood-brain barrier penetration or activity of biological targets. Corwin Hansch initiated the field of quantitative structure-activity relationships in the years 1962 and 1963, and they reported a study on the structure-activity relationships of plant growth regulators and their dependency on Hammett constants and hydrophobicity with the publications [1, 2].

The concept of QSAR is used for drug discovery and development and has gained wide applicability for correlating molecular information with biological activities, and the quantitative structure-property relationship (QSPR) is an alternative to experimental processing that envisages various physical and chemical properties. QSPR is related to the structure and any physical-chemical properties of the compounds taken into account. QSAR/QSPR associates biological activities or physical-chemical properties with certain structural features or atomic, group or molecular descriptors in the series of compounds. The QSAR/QSPR model includes structure representation, descriptive analysis and modeling. Todeschini and Consonni [3] defined the molecular descriptor as the following "The molecular descriptor is the final result of a logic and mathematical procedure which transforms chemical information encoded within a symbolic representation of a molecule into a useful number or the result of some standardized experiment." Chemical structural features are called molecular descriptor, and they are closely related to target property of the compounds. There are many molecular descriptors. Some of them are conformational, fragment constants, electronic, receptor, quantum mechanical, graph-theoretic, topological, information-content, molecular shape analysis, spatial, structural, thermodynamic, pKa, Absorption, distribution, metabolism, and excretion (ADME), molecular field analysis and receptor surface analysis descriptors. The descriptors may be classified as topological, geometrical, electronic and hybrid or 3D descriptors.

Topological indices are two-dimensional descriptors which take into account the internal atomic arrangement of compounds, and which encode in numerical form information about molecular size, shape, branching, presence of heteroatoms and multiple bonds and are a very useful tool for drug design specialists, with advantages such as offering a simple way of measuring molecular branching, shape and size [4, 5]. Third generation of topological indices is the hyper-Wiener index [6, 7] or the molecular identification (ID) numbers [8], the information indices [9–11], and the electrotopological state (E-state) indices [12, 13].

Geometrical descriptors or 3D descriptors in general provide much more information and discrimination power than topological descriptors for similar molecular structures and molecule conformations due to involving knowledge of the relative positions of the atoms in 3D space [14].

A number of geometric descriptors have been proposed by several scientific communities in the last decade to get molecular information for development of QSAR/QSPR models [3].

Electronic descriptors can be used in the design of a training set in QSAR studies, and the electronic identifiers obtained by quantum mechanical calculations are more precisely than those obtained by semiempirical calculations [15].

Quantum chemically derived descriptors can be subdivided as atomic charges, molecular orbital energies, frontier orbital densities, atom-atom polarizabilities, molecular polarizability, dipole moment and polarity indices, and energy [16], free valence of atoms [17], atomic orbital electron populations [18], overlap populations [19], partitioning of energy data into one-center and two-center terms [19], and vectors of lone pair densities [19] are the other quantum chemical descriptors successfully used in QSAR/QSPR studies.

Since electrostatic interactions play important role in a chemical reaction, one of the most fundamental descriptors to be used in QSAR are quantum chemically computed atomic charges. The atomic charges have been used for the prediction of anti-HIV-1 activities of 1-[(2-hydroxyethoxy)methyl]-6-(phenylthio)thymine (HEPT)-analog compounds [20]. They explained octanol-water partition coefficients of organic compounds with the atomic charges [16, 21]. Bhat et al. [22] reported optimal ligand-charge distribution at protein-binding sites with the help of atomic charge

Highest occupied molecular orbital (HOMO) and lowest unoccupied molecular orbital (LUMO) are very popular quantum chemical descriptors. The strongest Frontier orbitals (FO)

interaction involves the HOMO of the nucleophile and the LUMO of the substrate [23]. They reported that mutagens have lower LUMO energies than nonmutagens [24] and also reported that carcinogens, as a group, have lower LUMO energies than noncarcinogens [25].

As a conclusion, a QSAR/QSPR tries to find a consistent relationship between molecular properties and variability in biological activity for a number of compounds so that these equations can be used to evaluate new chemical entities.

QSAR has been applied successfully and extensively to find predictive models for activity of bioactive agents for the toxicity prediction [26–29], activity of peptides [30–33], drug metabolism [34–36], gastrointestinal absorption [37–39], prediction of pharmacokinetic and ADME properties [40–44], drug resistance and physicochemical properties [45–47].

Author details

Fatma Kandemirli

Address all correspondence to: fkandemirli@yahoo.com

Department of Biomedical Engineering, Faculty of Engineering and Architecture, University of Kastamonu, Kastamonu, Turkey

References

[1] Hansch C, Maloney PP, Fujita T, Muir RM. Correlation of biological activity of phenoxyacetic acids with Hammett substituent constants and partition coefficients. Nature. 1962;**194**:178-180

[2] Hansch C, Muir RM, Fujita T, Maloney PP, Geiger CF, Streich M. The correlation of biological activity of plant growth-regulators and chloromycetin derivatives with Hammett constants and partition coefficients. Journal of the American Chemical Society. 1963;**85**:2817-2824

[3] Todeschini R, Consonni V. Handbook of Molecular Descriptors. Wiley-VCH; Weinheim, 2000

[4] Gozalbes R, Doucet JP, Derouin F. Application of topological descriptors in QSAR and drug design: History and new trends. Current Drug Targets—Infectious Disorders. 2002;**2**:93-102

[5] Ivanciuc O, Balaban AT. In: Devillers J, Balaban AT, editors. Topological Indices and Related Descriptors in QSAR and QSPR. The Netherlands: Gordon and Breach Science Publishers; 1999. pp. 59-167

[6] Gutman I. A formula for the Wiener number of trees and its extension to graphs containing cycles, Graph Theory Notes, New York. 1994;**27**:9-15

[7] Randié M, Guo X, Oxley T, Krishnapriyan H. Wiener matrix: Source of novel graph invariants. Journal of Chemical Information and Computer Sciences. 1993;**33**:709-716

[8] Ivanciuc O, Balaban AT. Design of topological indices. Part 3. New identification numbers for chemical structures: MINID and MINSID. Croatica Chemica Acta. 1996;**69**:9-16

[9] Balaban AT, Balaban TS. New vertex invariants and topological indices of chemical graphs based on information on distances. The Journal of Mathematical Chemistry. 1991;**8**:383-397

[10] Carter S, Trinajstić N, Nikolić S. A note on the use of ID numbers in QSAR studies. Acta Pharmaceutica Jugoslavica. 1967;**37**:37-42

[11] Carter S, Trinajstić N, Nikolić S. On the use of ID numbers in drug research: A QSAR of neuroleptic pharmacophores. Medical Science Research. 1988;**16**:185-186

[12] Kier LB, Hall LH. An electrotopological-state index for atoms in molecules. Pharmaceutical Research. 1990;**7**:801-807

[13] Hall LH, Kier LB. Electrotopological state indices for atom types: A novel combination of electronic, topological, and valence state information. Journal of Chemical Information and Computer Sciences. 1995;**35**:1039-1045

[14] Tomasz P, Jerzy L, Mark TC. Recent Advances in QSAR Studies: Methods and Applications. Dordrecht, Heidelberg, London, New York: Springer 2010

[15] Cartier A, Rivail JL. Electronic descriptors in quantitative structure activity relationships. Chemometrics and Intelligent Laboratory Systems. 1937;**1**(4):335-347

[16] Karelson M, Lobanov VS, Katritzky AR. Quantum-chemical descriptors in QSAR/QSPR studies. Chemical Reviews. 1996;**96**(3):1027-1044

[17] Prabhakar YS. Quantum QSAR of the antirhinoviral activity of 9-benzylpurines. Drug Design and Delivery. 1991;**7**:227-239

[18] Cardozo MG, Iimura Y, Sugimoto H, Yamanishi Y, Hopfinger AJ. QSAR analyses of the substituted indanone and benzylpiperidine rings of a series of indanone-benzylpiperidine inhibitors of acetylcholinesterase. Journal of Medicinal Chemistry. 1992;**35**:584-589

[19] Sklenar H, Jager J. Molecular structure–biological activity relationships on the basis of quantum-chemical calculations. International Journal of Quantum Chemistry. 1979;**16**:467-484

[20] Alves CN, Pinheiro JC, Camargo AJ, Ferreira MMC, da Silva ABF. A structure–activity relationship study of HEPT-analog compounds with anti-HIV activity. Journal of Molecular Structure (THEOCHEM). 2000;**530**:39-47

[21] Ghose AK, Pritchett A, Crippen GM. Atomic physicochemical parameters for three dimensional structure directed quantitative structure-activity relationships III: Modeling hydrophobic interactions. Journal of Computational Chemistry. 1988;**9**:80-90

[22] Bhat S, Sulea T, Purisima EO. Coupled atomic charge selectivity for optimal ligand charge distributions at protein binding sites. Journal of Computational Chemistry. 2006;**27**:1899-1907

[23] Nguyễn TA. Frontier Orbitals: A Practical Manual. John Wiley & Sons Ltd.; Southport, Merseyside, United Kingdom 2007

[24] Rosenkranz HS, Klopman G. Decreased electrophilicity of chemical carcinogenic only at the maximum tolerated dose. Mutation Research. 1992;**282**(4):241-246

[25] Rosenkranz HS, Klopman G. Relationships between electronegativity and genotoxicity. Mutation Research. 1995;**328**:215-227

[26] Patricia R, Gino B, Terry T, John W, Moiz M. Prediction of acute mammalian toxicity using QSAR methods: A case study of sulfur mustard and its breakdown products molecules. 2012;**17**:8982-9001

[27] Demchuk E, Ruiz P, Chou S, Fowler BA. SAR/QSAR methods in public health practice. Toxicology and Applied Pharmacology. 2011;**254**:192-197

[28] Ruiz P, Mumtaz M, Gombar V. Assessing the toxic effects of ethylene glycol ethers using quantitative structure toxicity relationship models. Toxicology and Applied Pharmacology. 2011;**254**:198-205

[29] Enslein K. The future of toxicity prediction with QSAR. In Vitro Toxicology. 1993;**6**: 163-169

[30] Tong JB, Chang J, Liu SL, Bai M. A quantitative structure–activity relationship (QSAR) study of peptide drugs based on a new descriptor of amino acids. Journal of the Serbian Chemical Society. 2015;**80**(3):343-353

[31] Mariya AT, Aleksandar MV, Jovana BV, Dušica BS, Andrey AT. QSAR modeling of the antimicrobial activity of peptides as a mathematical function of a sequence of amino acids. Computational Biology and Chemistry. 2015;**59**:126-130

[32] Mikut R, Hilpert K. Interpretable features for the activity prediction of short antimicrobial peptides using fuzzy logic. International Journal of Peptide Research and Therapeutics. 2009;**15**(2):129-137

[33] Reyhaneh J, Somaieh S, Abolfazl B. A review of QSAR studies to predict activity of ACE peptide inhibitors. Pharmaceutical Sciences. 2014;**20**:122-129

[34] Wenlock MC, Carlsson LA. How experimental errors influence drug metabolism and pharmacokinetic QSAR/QSPR models. Journal of Chemical Information and Modeling. 2015;**55**:125-134

[35] Klopman G, Dimayuga M, Talafous J. META. 1. A program for the evaluation of metabolic transformation of chemicals. Journal of Chemical Information and Modeling. 1994;**34**:1320-1325

[36] Braga RC, Andrade CH. QSAR and QM/MM approaches applied to drug metabolism prediction. Mini-Reviews in Medicinal Chemistry. 2012;**12**:573-582

[37] Yuan HZ, Michael HA, Joelle L, Anne H, Chris NL, Gordon B, Brad S, Ian C. Rate-limited steps of human oral absorption and QSAR studies. Pharmaceutical Research. 2002;**19**(10):1446-1457

[38] Zhao YH, Le J, Abraham MH, Hersey A, Eddershaw PJ, Luscombe CN, Butina D, Beck G, Sherborne B, Cooper I, Platts JA. Evaluation of human intestinal absorption data for use in QSAR studies and a quantitative relationship obtained with the Abraham descriptors. Journal of Pharmaceutical Sciences. 2001;**90**:749-784

[39] Gabriele C, Emanuele C, De Benoit B, Kantharaj E, Claire M, Trevor H, Riccardo V. MetaSite: Understanding metabolism in human cytochromes from the perspective of the chemist. Journal of Medicinal Chemistry. 2005;**48**:6970-6979

[40] Majid Z, Mohammad S, Farzin H, Mohammad AD, Kaveh T. Prediction of pharmacokinetic parameters using a genetic algorithm combined with an artificial neural network for a series of alkaloid drugs. Scientia Pharmaceutica. 2014;**82**:53-70

[41] Norris DA, Leesman GD, Sinko PJ, Grass GM. Development of predictive pharmacokinetic simulation models for drug discovery. Journal of Controlled Release. 2000;**65**:55-62

[42] Pires DEV, Blundell TL, Ascher DB. pkCSM: Predicting small-molecule pharmacokinetic and toxicity properties using graph-based signatures Journal of Medicinal Chemistry. 2015;**58**:4066-4072

[43] Cao D, Wang J, Zhou R, Li Y, Yu H, Hou T. ADMET evaluation in drug discovery. 11. PharmacoKinetics Knowledge Base (PKKB): A comprehensive database of pharmacokinetic and toxic properties for drugs. Journal of Chemical Information and Modeling. 2012;**52**:1132-1137

[44] Obach RS, Lombardo F, Waters NJ. Trend analysis of a database of intravenous pharmacokinetic parameters in humans for 670 drug compounds. Drug Metabolism and Disposition. 2008;**36**:1385-1405

[45] Narender S, Sidhartha C, Ruifeng L, Mohamed Diwan MA, Gregory T, Anders W. QSAR classification model for antibacterial compounds and its use in virtual screening. Journal of Chemical Information and Modeling. 2012;**52**:2559-2569

[46] Hu Y, Unwalla R, Denny RA, Bikker J, Di L, Humblet C. Development of QSAR models for microsomal stability: Identification of good and bad structural features for rat, human and mouse microsomal stability. Journal of Computer-Aided Molecular Design. 2010;**24**:23-35

[47] Bray PG, Hawley SR, Mungthin M, Ward SA. Physicochemical properties correlated with drug resistance and the reversal of drug resistance in *Plasmodium falciparum*. Molecular Pharmacology. 1996;**50**(6):1559-1566

Chapter 2

Computational Approaches in the Development of Phosphodiesterase Inhibitors

Anand Gaurav, Ming Xing and Mayasah Al-Nema

Additional information is available at the end of the chapter

http://dx.doi.org/10.5772/intechopen.68842

Abstract

Computational drug design tools have become indispensable in the quest for new drugs. There is hardly any drug discovery program where computational methods are not employed, be it structure-based or ligand-based methods. Numerous drug targets have been explored for discovery of new drugs using computational methods. In recent times, discovery of newer and selective phosphodiesterase as medications for inflammatory disorders, CNS disorders, and many other diseases has been the focus of many research groups worldwide. Most of these groups have employed computational methods of drug design and discovery at different stages of their research. This chapter reviews the reported application of computational methods used in the discovery and development of phosphodiesterase inhibitors.

Keywords: phosphodiesterase, QSAR, docking, molecular dynamics, pharmacophore

1. Introduction

The early application of computational methods as a means to identify and design phosphodiesterase (PDE) inhibitors goes back to the 1980s where the first reported study tried to explain the relation between specific physicochemical properties and potency of known inhibitors [1]. The result was the identification of pharmacophore for PDE inhibitors [2]. Computational methods of drug design and discovery have impacted the overall process of drug discovery in a big way over the last two decades [3, 4]. The cost as well as time of drug discovery has been reduced due to the inclusion of computational methods as well as high-throughput synthesis and screening. Computational methods have undergone development at immense pace and newer and more accurate methods have come up. Traditionally, computational methods have been classified into structure-based methods and ligands-based methods [4]. Structure-based methods like docking-based virtual screening has been used for new lead identification for phosphodiesterase (PDE) inhibitors, while ligand-based methods such as quantitative structure activity relationships (QSAR) have been mostly used for lead optimization [5–8]. However, ligand-based methods, such as pharmacophore development, have played an important role in many lead discovery programs. Structure-based methods like molecular dynamics (MD) simulations have been used to study the binding of PDE inhibitors with the enzyme. While the information derived from MD simulations has been used to optimize the structure of inhibitors [9–13]. As far as exploring the inhibitor structural features determining selectivity variations, the computational approach could rely on both structure-based and ligand-based strategies. However, due to high degree of similarity observed among all the PDE catalytic sites, the ligand-based approach could represent a much more promising tool to deeply investigate the selectivity issues around PDE inhibitors as compared to structure-based approach.

This chapter aims to provide an update of the recent advances in the field of rational design of PDE inhibitors. Attempt has been made to explore both scientific journals as well as patent literature.

2. Ligand-based methods: general aspects and perspective

The most prominently used ligand-based drug design methods include QSAR and pharmacophore mapping. There are several reports where QSAR has been used for optimizing the PDE inhibition activity of ligands having specific structural scaffolds. Some studies have reported unique applications of ligand-based methods, i.e., ligand-based homology modeling and exploration of selectivity toward specific receptor subtypes.

2.1. PDE10A inhibitors

Among the several PDE subtypes inhibitors, PDE10A inhibitors have attracted enormous interest recently for their potential in the treatment of schizophrenia and Huntington's disease, whereby they can fill up the voids present in the current therapeutic approach. In one of the recent studies on PDE10A inhibitors, Mondal et al. reported the application of ligands-based methods toward optimization of cinnolines as PDE10A inhibitors [14]. 2D-QSAR, HQSAR (hologram quantitative structure activity relationship), pharmacophore mapping, and three-dimensional (3D)–QSAR analyses in combination with structure-based methods like molecular docking and MD simulations were used for the purpose.

Eighty-one cinnoline derivatives having PDE10A inhibitory activity were used as the data set. 2D-QSAR models were developed by multiple linear regression and partial least square (PLS) analyses using both atom-based and whole molecular descriptors. The best model, having considerable internal ($q^2 = 0.812$) and external (R^2 pred = 0.691) predictabilities, demonstrated importance of atom-based topological and whole molecular E-state as well as 3D topological indices.

Hologram QSAR, a relatively newly developed QSAR technique, relates biological activity to structural fragments [15]. HQSAR eliminates the need for generation of 3D structures, putative binding conformations, and molecular alignments. The process of generating HQSAR models involve the fragmentation of the data set structures and then hashing into array bins. Molecular hologram fingerprints are then generated. Holograms are constructed by cutting the fingerprint into strings at various hologram length parameters. After the generation of descriptors, partial least square (PLS) methodology is used to find the possible correlation between dependent variable (activity) and independent variable (descriptors generated by HQSAR structural features) [15]. The best HQSAR model for cinnolines as PDE10A inhibitors was found to be statistically significant ($q^2 = 0.664$, R^2 pred = 0.513), and it highlighted some important structural features.

Pharmacophore mapping was also employed in this case. The pharmacophore hypothesis showed the importance of hydrogen bond acceptors and ring aromatic and hydrophobic features for higher activity. The compounds were mapped to the pharmacophore as per their activity, and the pharmacophore provided an efficient means of aligning the compounds for comparative molecular field analysis (CoMFA) and comparative molecular similarity indices analysis (CoMSIA) studies.

Finally, three-dimensional QSAR methods implemented as CoMFA and CoMSIA were applied on the data set. 3D-QSAR models were generated using two different types of alignment procedures—(1) pharmacophore based and (2) docking based. Docking-based alignment produced better results for both CoMFA ($q^2 = 0.578$; R^2 pred = 0.841) and CoMSIA ($q^2 = 0.610$; R^2 pred = 0.824) methods.

Molecular dynamics (MDs) simulations were also performed for two ligand–receptor complexes, and the findings of MD simulations were consistent with the interpretations obtained from ligand-based analyses methods described above. The role of hydrogen bond acceptors and aromatic and hydrophobic features in determining PDE10A inhibition potential was unraveled. The study clearly puts forth guideline for the design and development of cinnolines as potent and selective PDE10A inhibitors and demonstrates the role of ligand-based drug design methods combined with structure-based drug design (SBDD) methods.

2.2. PDE4 inhibitors

PDE4 has long been considered as target for design of antiinflammatory molecules, and numerous PDE4 inhibitors have been developed [16–21]. Quinolines as a third-generation of PDE4 inhibitors show an increased antiinflammatory effect without being dose limited by side effects compared to the first- and second-generation of PDE4 inhibitors such as rolipram and roflumilast. Recently, newer quinoline derivatives were developed as selective PDE4B inhibitors using ligand-based pharmacophore and atom-based 3D-QSAR modeling along with structure-based docking and ADME methods [21].

These studies, in combination, led to new understanding about the selection of target molecules from many candidates. In this case, computational methods were not the primary tools used for designing the ligands rather they were used for identifying and confirming the binding site of the ligands on PDE4B. The 3D-QSAR model of PDE4B inhibitors developed for this study proved to be reliable with r^2 value of 0.96 and q^2 value of 0.91. The specific pharmacophore for PDE4B was then mapped and selected for virtual screening, and the potent PDE4B inhibitors were finally confirmed their selectivity ability for PDE4B by docking, ADME analysis, and MD, and then, molecules were developed as selective PDE4B inhibitors (**Figure 1**) [21].

2.3. PDE7 inhibitors

Like PDE4, PDE7, a cAMP-specific phosphodiesterase, is also highly expressed in human immune system such as thymus, lymph nodes, spleen, and blood leukocytes, suggesting PDE7 as a possible target for treating CNS and airway diseases [22–25]. A recent study by Cichero et al. whereby they applied 3D-QSAR methods for the development of selective PDE 7 inhibitors is worth noting. Herein, 72 derivatives of thieno[3,2-d]pyrimidin-4(3H)-one were selected from literature and were used to develop new selective PDE7 inhibitors [26]. The uniqueness of the study was the application of 3D-QSAR to identify the structural features required for the selectivity for a particular target.

Docking-based 3D-QSAR followed by redocking was performed to identify the most suitable bioactive conformations of derivatives and most comparable binding mode with that of X-ray structure. The conformations of the compounds showed good agreement with the conformations in the known PDE7A-ligand complex. The compounds were further submitted to a CoMFA and CoMSIA analysis. Since inhibitors which showed high selectivity for PDE7, rather than PDE4, are candidates for antiinflammatory molecules, two models (PDE7 selectivity (model A) and

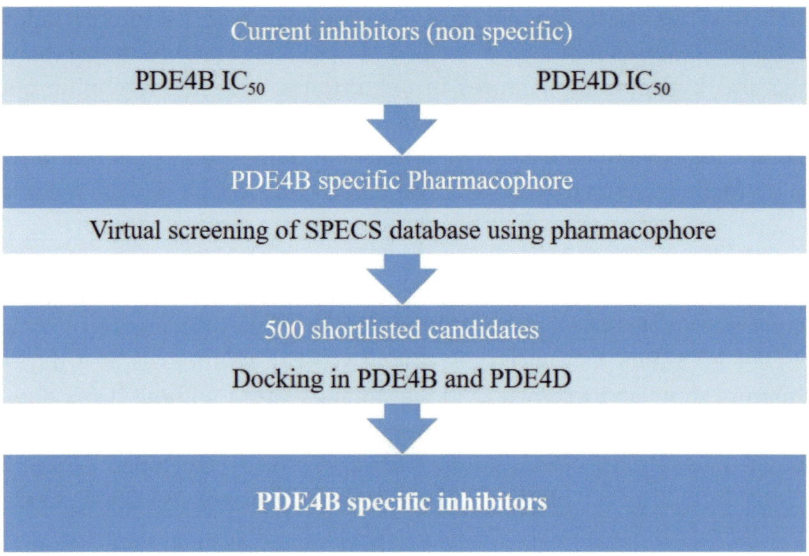

Figure 1. *In silico* strategy to discover PDE4B specific inhibitors.

PDE4B selectivity (model B)) were established. Predictive models for both the activities were obtained, i.e., CoMFA$_{\text{(model A)}}$ (optimum number of components (ONC = 6, r^2 = 0.946, and standard error of estimate, SEE, = 0.292), CoMSIA$_{\text{(model A)}}$ (r^2 = 0.961, SEE = 0.249), CoMFA$_{\text{(model B)}}$ (ONC = 6, r^2 = 0.968, and SEE = 0.353), and CoMSIA$_{\text{(model B)}}$ (r^2 = 0.968 and SEE = 0.356; **Figure 2**). The role of steric, electrostatic, and hydrophobic features in the binding mechanism of ligands to both PDE7 and PDE4 was identified. The study unraveled structural information for the development of new PDE7 inhibitors with high selectivity.

2.4. PDE11 inhibitors

Another unique application of ligand-based methods was reported by Cichero et al. earlier in 2013. Since the X-ray structure of PDE11 was not available, ligand-based homology modeling

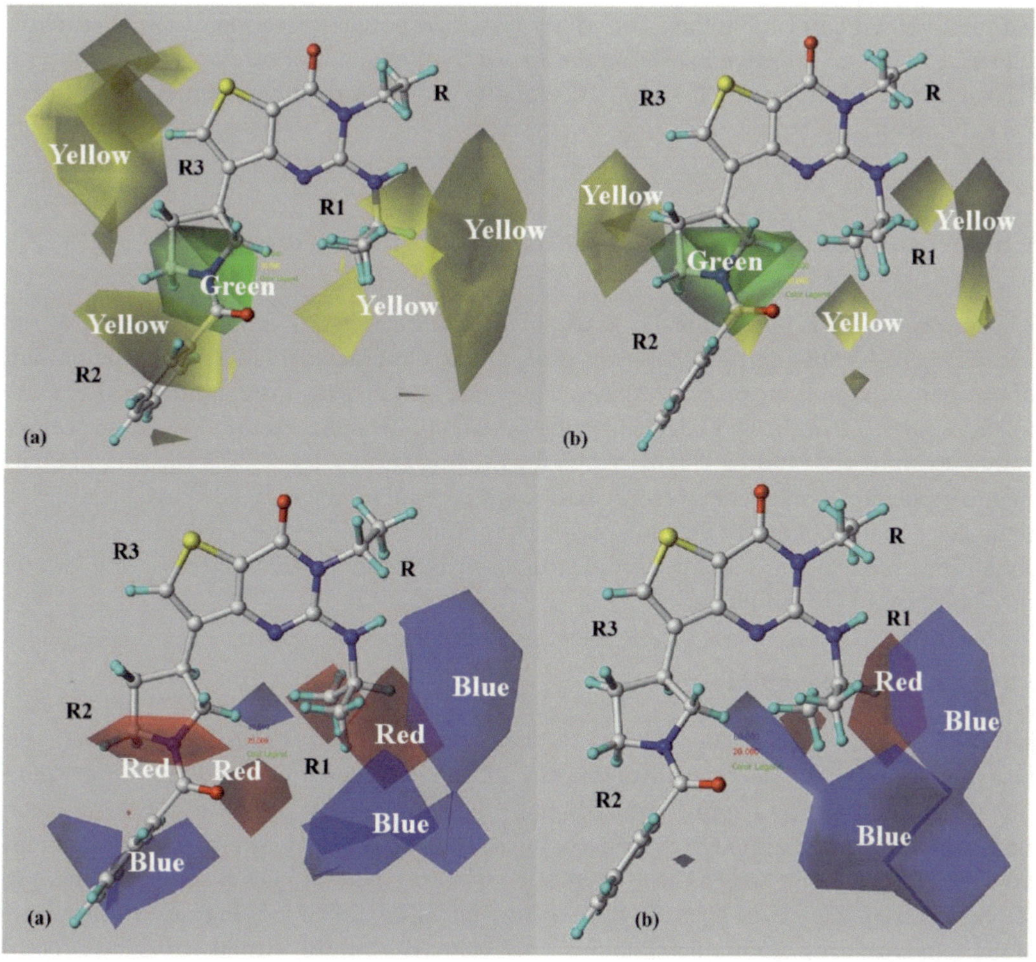

Figure 2. CoMFA and CoMSIA models for (a) PDE7 selectivity and (b) PDE4B selectivity (CoMFA: green, favored for bulky groups; yellow, disfavored for bulky groups) (CoMSIA: blue, favorable for more positively charged groups; red, favorable for less positively charged groups). (Cichero et al. [26] – Published by The Royal Society of Chemistry).

technique was applied to explore the 3D structure of PDE11 [27]. Consistent with the sequence similarity of PDE11 and PDE5, several tadalafil analogs (PDE5 ligands) showed binding affinity to PDE11 too [28–30]. Thus, the 3D structure of PDE11 was built on the basis of PDE5-tadalafil complex using homology modeling technique [31]. Besides the conventional homology modeling steps such as insertion and deletion of extra atoms during the energy minimization stages, evolution, including model building and refinement were performed in this specific case. Specifically, residues located in H-loop and M-loop were involved in obtaining PDE11 structure as they are important for substrate recognition [32]. The coordinates of PDE11 were derived from PDE5-tadalafil complex, and the amino acid sequences were obtained from the SWISSPROT database. This was followed by the development of models using MOE software without significant main chain deviations. AMBER94 force field was applied to minimize the structure energy, and Ramachandran plots were applied to assess the final obtained model. Successively, reliability of the derived PDE11 model was further confirmed by molecular docking and MD simulations with selective PDE11 inhibitors and dual PDE5-PDE11 inhibitors. This study provided new information of target structure of PDE11 to design selective PDE11 inhibitors for the treatment of cardiac pathologies [33]. The success of this approach is supposedly due to the synergic interaction between theory and experiment.

3. Structure-based methods: general aspects and perspective

Structure-based drug design (SBDD), as one of the *in silico* methods, is almost an integral part of any drug discovery and development project. These methods, in addition to the binding affinity between a specific protein target and a ligand, also provide insight into the interaction between the two. This helps in devising substituent modifications around the ligand scaffold leading to improved binding [34]. Based on the existing knowledge of 3D structures of PDEs, the great potential and success of structure-based computational methods have been visible in the development of newer PDE inhibitors. In addition to the routine screening and interaction studies, many unique applications of structure-based methods have been reported for the development of PDE inhibitors.

3.1. Selective PDE4B inhibitors

In one of the recent reports by Jing Li and team, virtual screening of natural product database was carried out to discover novel selective PDE4B inhibitors from natural products database [35]. Structure-based approaches like docking and molecular dynamics simulation were used for the purpose, although pharmacophore-based screening was also employed as initial filter. The screening led to the identification of four potential PDE4B-selective inhibitors (ZINC67912770, ZINC67912780, ZINC72320169, and ZINC28882432; **Figure 3**). Compared to the reference drug (roflumilast), they scored better during the virtual screening process. DOCK and Vina were used for docking, and results were agreeable. Binding free energy was −317.51, −239.44, −215.52, and −165.77 kJ/mol, which is better than −129.05 kJ/mol of roflumilast. The MD studies also showed that ZINC28882432-PDE4B complex reached stable RMSD

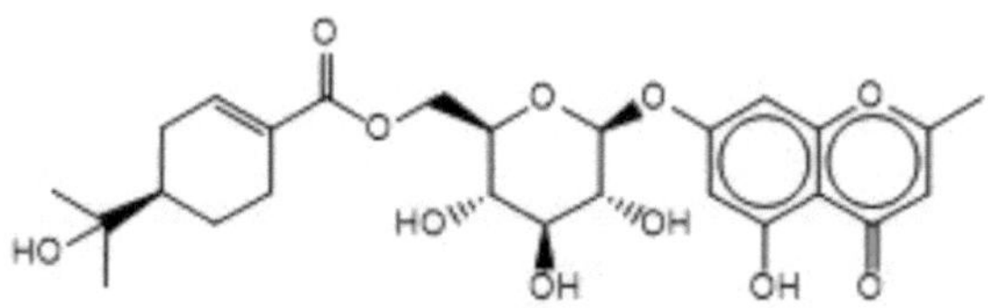

ZINC28882432

Figure 3. Structure of ZINC28882432 a PDE4B inhibitor from natural sources identified by virtual screening.

faster than roflumilast-PDE4B complex. This study demonstrates the successful application of MD in a virtual screening workflow.

3.2. Selective PDE5 inhibitors

The cross-reactivity of PDE 5 inhibitors like sildenafil, vardenafil, and tadalafil with hERG1 is well documented. However, hERG1 is proven to be involved in the regulation of human ventricular myocyte action potential, thus blocking this hERG channel results in the function loss of the PDE5 inhibitors and further leads to serious life-threatening disorders and cardiovascular problems [36, 37]. One of the recent studies explored this cross-reactivity using docking. Open and open-inactivated states of hERG1 potassium channel were used as protein structures and binding interaction patterns and affinity between the proteins and PDE5 inhibitors were studied [38]. Three different structure-based docking tools including GOLD, MOE, and AUTODOCK were applied to identify the binding interactions between Sildenafil and hERG1 channel. In open-state conformation of hERG1, both GOLD and MOE showed common interactions and further AUTODOCK and GOLD scoring analysis gave a 2.16 kcal/mol lower energy compared to open-inactivated state. In open-inactivated conformation of hERG1, three docking tools showed similar interaction patterns but MOE docking results were more specific in terms of H-bonding, distance, and key residues. To further confirm the key residues which are responsible for binding affinity to the hERG1 channel, *in silico* alanine mutagenesis study was performed, and new promising molecules were designed based on the interaction patterns and alanine mutagenesis studies. MD simulations were finally used to confirm that the complex of the new compounds with hERG1 channel is much less stable than that with PDE5. This study showed new approach toward the design of PDE5 inhibitors with lower affinity for hERG1 channels.

In another study dealing with cross-reactivity of PDE5 inhibitors, Kayık and team explored cross-reactivity of PDE5 inhibitors with PDE6 and PDE11 [39]. The major challenge of their study was designing novel PDE5 inhibitors with decreased cross-reactivity with PDE6 and PDE11. For this aim, the similarity-based virtual screening protocol was applied for the "clean drug-like subset of ZINC database that contained more than 20 million small compounds. Moreover, molecular dynamics (MD) simulations of selected hits complexed with PDE5 and off-targets were performed to get insights for structural and dynamical behaviors of the selected molecules as selective PDE5 inhibitors. Since tadalafil blocks hERG1 K channels in concentration-dependent manner, the cardiotoxicity prediction of the hit molecules was also tested. The study revealed important structural information for the design of novel, safe, and selective PDE5 inhibitors by applying SBDD.

3.3. Dual PDE1 and PDE5 inhibitors

Dual inhibitors of PDE1 and PDE 5 are known to produce vasodilation and have thus been explored as therapeutic agents for the treatment of different cardiovascular diseases, i.e., hypertension, angina, heart failure, and arteriosclerosis [40, 41]. Yamazaki et al. employed a ligand-based virtual screening for the identification of novel lead candidates with potent dual inhibition of PDE1 and 5 [42]. These compounds have application in the treatment of different cardiovascular diseases, i.e., hypertension, angina, heart failure, and arteriosclerosis [42, 43].

They applied virtual screening that consists of classification and regression tree (CART) analysis with the utilization of 168 2-center pharmacophore descriptors and 12 macroscopic descriptors which can result in finding of new lead compounds and drug candidates efficiently. The method applied started with the learning step of CART analysis where a prediction model is configured, and the explanatory variables are selected as per the discrimination index of training set. These variables should be independent from each other.

In the following step, they checked the potential energy term and solvation energy, which are the main constituents of the binding energy used in molecular modeling and simulation, to select the most suitable set of explanatory variables. The potential and solvation energy was obtained by summing up electrostatic potential term, van der Waals potential term, solvation free energy term, and hydrogen bond energy term. This situation is represented qualitatively by pharmacophore where it provides the binding features with the position of these features. In chemoinformatic analysis, the pharmacophore descriptor is either a part or all of pharmacophore. The *n*-center pharmacophore descriptor is the number of features that participate in binding and the distance between them, for example, 2-center, 3-center, etc. In this research, they adapted 2 for *n* to reduce the number of explanatory variables. Six binding features were used with eight classes of distance between them, this gives 168 two-center pharmacophore descriptor.

In next stage, the virtual screening has been performed to screen the library of commercially available chemical compounds that was supplied by SPECS Inc. for PDE5 inhibitor activity. Based on Lipinski's rule of five, the compounds with undesirable physicochemical and pharmacokinetic properties were filtered out, followed by selection of compounds with desirable inhibition activities for PDE5 by CART model. The next step involved the selection of structurally diverse 100 compounds. Finally, 19 drug-like compounds out of 100 were selected and obtained from SPECS Inc. and assayed *in vitro* to test their inhibitory activity against PDE1 and PDE5 [42]. The results showed that the virtual screening along with the CART analysis have a high prediction capability for biological activity of new chemical compounds.

3.4. Selective PDE1 inhibitors

One of the least explored PDEs as a drug target singly has been PDE1, although it has been a combined target with PDE5 for vasodilator drugs [11, 44]. However, some of the studies have shown that it is a promising target for the treatment of cognitive impairments [45]. One such report by Li et al. recently presented novel PDE1 inhibitors for the treatment of cognitive impairments. They applied both ligand- and structure-based drug design methods to discover these novel PDE1 inhibitors. Achieving high selectivity among PDE enzymes is challenging, since all PDE enzymes share a high degree of sequence homology in the catalytic domains [46–50]. They started by analyzing the published PDE inhibitors with various scaffolds and found out that pyrazolo[3,4-d]pyrimidinones, which was reported by Xia and coworkers, could be the first step for designing a new, potent, and selective PDE1 inhibitors [51]. At the beginning, these compounds were developed as dual inhibitors for both PDE1 and PDE5 [46–49].

According to the available X-ray crystal structures of human PDE enzymes, a hydrogen bond is formed between the N-7 nitrogen of purine ring in guanine and adjacent amino acid residue in catalytic site. In pyrazolo[3,4-d]pyrimidinone compounds, the hydrogen bond network is disrupted due to the shift of nitrogen in pyrozole ring from position 3 to position 2. Li et al. designed polycyclic 3-aminopyrazolo[3,4-d]pyrimidinone scaffold 2 to reestablish the hydrogen bond through the substitution of amino group at carbon 3 of pyrazole ring [50]. By applying a combination of ligand-based and structure-based drug design methods, they designed numerous novel scaffolds as PDE1 inhibitors. Their work resulted in the discovery of a clinical candidate (ITI-214) which has excellent selectivity toward PDE1 (**Figure 4**). ITI-214 is now in clinical trial phase I and is being tested for its ability in the treatment of cognitive deficits associated with neurodegenerative and neuropsychiatric disorders and other CNS and non-CNS diseases [45, 50].

3.5. PDE9A inhibitors

Another PDE which has emerged as a promising drug target recently is PDE9A [52]. PDE9A is now considered as an important therapeutic target for the treatment of diabetes and Alzheimer's disease (AD) [53]. Most of the inhibitors of PDE9A until recently were based on the pyrazolopyrimidinone scaffold, thus there was need to identify novel scaffolds possessing this activity [54]. Zhe Li and team used a combinatorial method including pharmacophores, molecular docking, molecular dynamics simulations, binding free energy calculations, and bioassay to discover novel PDE9A inhibitors with new scaffolds (**Figure 3**) [55]. SPECS database containing about

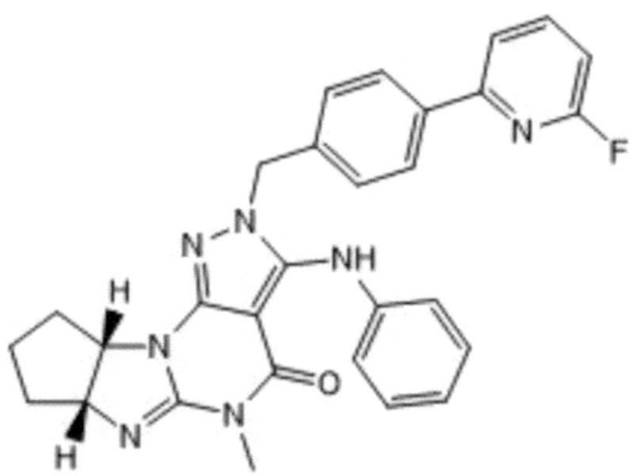

ITI-214

Figure 4. Structure of ITI-214, a PDE1 inhibitor in clinical trials.

200,000 compounds was screened using their combinatorial approach. The combination of ligand- and structure-based methods in combinatorial fashion was done with the aim to reduce computational cost as structure-based methods alone are computationally too expensive for virtual screening. The results were encouraging as has been the case whenever such combinatorial approach has been employed. Fifteen hits out of 29 molecules (a hit rate of 52%) with five novel scaffolds were identified to be PDE9A inhibitors with inhibitory affinities no more than 50 mM to enrich the structural diversity, different from the pyrazolopyrimidinone-derived family. The high hit ratio of 52% for this virtual screening method indicated that the combinatorial method is a good compromise between computational cost and accuracy. Binding pattern analyses indicate that those hits with nonpyrazolopyrimidinone scaffolds can bind the same active site pocket of PDE9A as classical PDE9A inhibitors. The five novel scaffolds discovered in this study can be used for the rational design of PDE9A inhibitors with higher affinities (**Figure 5**).

3.6. PDE2 inhibitors

PDE2 is a key enzyme that hydrolyzes both cAMP and cGMP. It has been suggested that selective PDE2 inhibitors could be a promising therapy for some of the CNS disorders, such as Alzheimer's disease, memory deficit, and depression since PDE2 modulates neuronal signaling

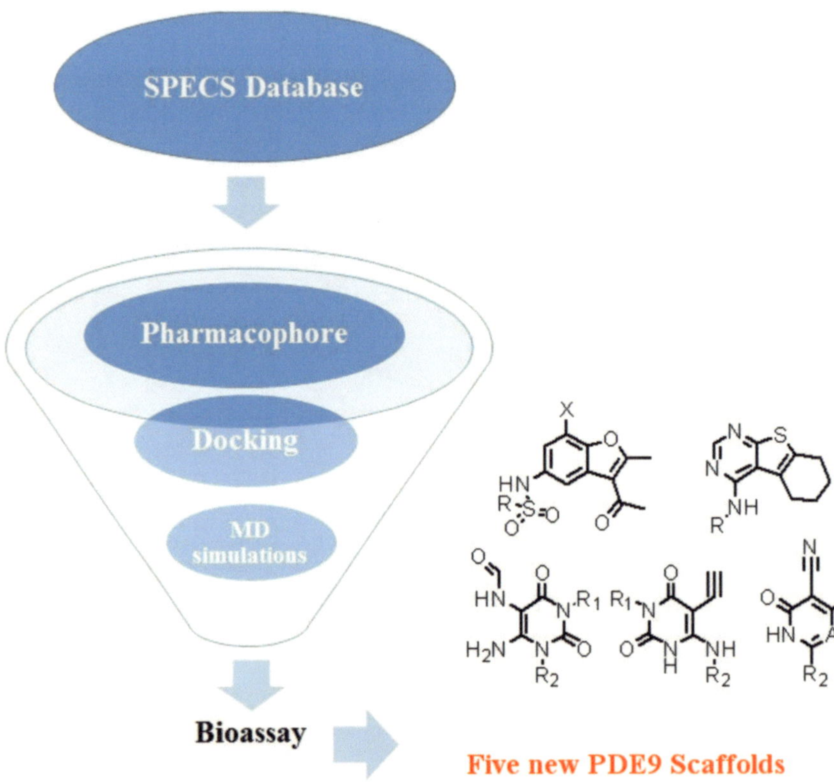

Figure 5. Virtual screening strategy to discover novel PDE9 inhibitor scaffolds.

involved in concentration, learning and memory, and emotion [56]. Bo Yang and coworkers examined the binding structures and free energies for PDE2 and benzo[1,4]diazepin-2-one derivatives (PDE2 inhibitors) by combining the molecular docking, molecular dynamics (MD), calculations of binding free energy, and binding energy decompositions [57, 58]. Molecular docking was performed followed by energy minimization. The docked structures were analyzed followed by the selection of the best pose for each ligand. The next step was MD stimulation. The binding free energy (ΔG_{bind}) of PDE2A-ligand complex was determined based on the MD trajectory for each complex [59]. The binding energy decomposition was estimated by using molecular mechanism/generalized Born surface area (MM/GBSA) method.

The results put forth important information regarding the PDE2A-ligand binding patterns including the intermolecular interactions, hydrophobic interactions, and hydrogen bonding. The estimated PDE2A-inhibitor binding patterns and the agreement between theoretical and experimental outcomes provided a firm base for further design of new, selective, and more potent PDE2A inhibitors [57].

In recent years, several potent inhibitors for PDE2 were developed but none reached the market due to either lack of selectivity or poor pharmacokinetic properties. This led work to focus on the optimization of pharmacokinetic properties of PDE2 inhibitors. Zhang et al. tried to discover selective potent PDE2A inhibitors with improved pharmacokinetic properties [60–67]. In their study, they described the identification of novel PDE2A inhibitors by structure-based virtual screening. They combined pharmacophore model-based screening and molecular docking along with MD simulations and bioassay to find new selective compounds with significant improvement in inhibition activity (**Figure 6**).

Beginning with the selection of small-molecule database SPECS (comprising almost 200,000 small molecules) for virtual screening, this database was filtered by Lipinski's rule of five to constitute databaset0 which is the initial data set. The crystal structures of PDE2A in a bound

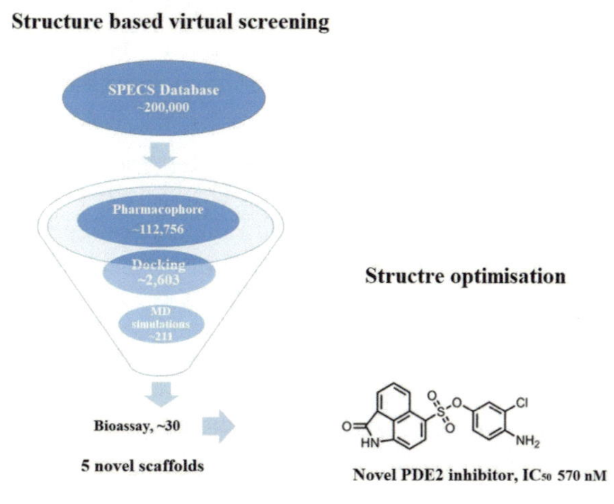

Figure 6. Discovery of LHB-8, a potent PDE2 inhibitor. (Reprinted with permission from Zhang et al. [67] Copyright © 2017 American Chemical Society).

state were used to generate the 3D pharmacophore model. This model was applied for minimizing the size of database through efficient screening of dataset0 to obtain dataset1. Pan-assay interfering compound substructures (PAINS) screening was used to obtain dataset2 by eliminating the false positive compounds that might interfere with other detecting methods and result in false positives.

In the next step, the dataset2 compounds were submitted for molecular docking to predict the preliminary optimal binding modes and binding energies between PDE2A and ligand. Those compounds with better scores than the reference and proper binding modes constituted dataset3. In the last step, MD stimulations and molecular mechanism/Poison-Boltzman surface area (MM-PBSA) method were applied for more accurate prediction of the binding modes and binding energies. AMBER 10.0 was utilized to stimulate the binding of PDE2A with the compounds in dataset3, whereas MM-PBSA was used for the calculation of binding free energy. At the end, 30 molecules with optimal binding patterns and top binding energies were selected to make up the final dataset. Nine hits out of 30 molecules (a hit rate of 30%) were identified with less than 50 μM affinity for PDE2A.

The result of this study was the discovery of new compound LHB-8 with IC_{50} = 570 nM (**Figure 4**). This compound poses novel scaffold benzo[cd]indol-2(1H)-one among PDE2A inhibitors, which can be used as the novel scaffold for designing of potent inhibitors of PDE2A [67].

3.7. PDE3 inhibitors

One of the early studies reporting the application of structure-based drug design on PDE3 inhibitors was done by Fossa et al. [2]. They combined homology modeling techniques with docking to gain knowledge about the molecular requirements of selective inhibition of PDE3 and identified the important amino acids residues for substrate and inhibitor binding. They built a homology model of PDE3A catalytic site based on coupling of PDE4B2B crystal structure and PDE3A mutagenesis data. The amino acids sequences of PDEs were obtained and multiple sequence alignments of PDE isozymes were performed [68–70]. Then, phylogenetic relationship was used to calculate the amino acid conservation based on the obtained alignments followed by prediction of secondary structure [71, 72]. Fold recognition and sequence to structure alignment was also done during the course [2, 73]. 3D models were generated by the application of restraint-based homology modeling methods [74]. At the end, the Amber force field was applied for energy minimization, and the generalized Born solvation model as implemented in AMBER was used for the calculation of molecular mechanic. The structural evaluation of the model proved that it is suitable for docking studies [75].

Another study was proposed by Kim et al. where they used a virtual screening approach to discover novel PDE3 inhibitors for obesity treatment. They started with the analysis of the structural features of the known PDE3 inhibitors and screening of virtual library with 30,000 diverse compounds, followed by docking study based on the 3D structure of PDE3B. In this work, Kim et al. built 3D structures of PDE3B-ligand complex by utilizing a cocrystal structure. Docking was carried out, and 80 compounds with low energy conformation were identified by docking. These compounds were examined for their adipocyte lipolysis activity and finally four structurally unrelated compounds were identified. Among these leads, the most potent compound has IC_{50} = 14.8 nM for PDE3A activity and 88.4 nM for PDE3B activity [76].

4. Future perspective

Apart from those discussed in the previous section, there are many more examples of successful application of *in silico* methods in the discovery and design of PDE inhibitors. The applications are very diverse ranging from lead discovery to optimization. However, the major problem with various PDE subtype inhibitors, designed or discovered, is the lack of selectivity. Although with the application of computational methods inhibitors with significant selectivity for PDE subtypes have been designed lately, still none of them have been able to make it to the market. The advancement in computational hardware and MD simulation methods will be the major boost for solving the selectivity problem in future. These advancements will allow more deeper studies on a tinier time scale allowing the computational chemist to capture interactions and structural features required for high degree of selectivity. These developments bring hope that the unfulfilled potential of various PDE inhibitors will be realized in near future.

Author details

Anand Gaurav[1]*, Ming Xing[2] and Mayasah Al-Nema[1]

*Address all correspondence to: anand.pharma@gmail.com

1 Faculty of Pharmaceutical Sciences, UCSI University, Kuala Lumpur, Federal Territory of Kuala Lumpur, Malaysia

2 Suzhou YHX Medical Technology Co. LTD, Suzhou, Jiangsu, China

References

[1] Blackmore, Timothy Robert. Monash University. Faculty of Pharmacy and Pharmaceutical Sciences. Medicinal Chemistry; 2012. Imidazoquinazolinone based inhibitors of Phosphodiesterase 3.

[2] Fossa P, Giordanetto F, Menozzi G, Mostia L. Structural basis for selective PDE 3 inhibition: A docking study. Quantitative Structure-Activity Relationships. 2002;**21**(3):267-275

[3] Sliwoski G, Kothiwale S, Meiler J, Lowe EW. Computational methods in drug discovery. Pharmacological Reviews. 2014;**66**(1):334-395

[4] Leelananda SP, Lindert S. Computational methods in drug discovery. Beilstein Journal of Organic Chemistry. 2016;**12**:2694-2718

[5] Kovalishyn V, Tanchuk V, Charochkina L, Semenuta I, Prokopenko V. Predictive QSAR modeling of phosphodiesterase 4 inhibitors. Journal of Molecular Graphics & Modelling. 2012;**32**:32-38

[6] Dong X, Ebalunode JO, Cho SJ, Zheng W. A novel structure-based multimode QSAR method affords predictive models for phosphodiesterase inhibitors. Journal of Chemical Information and Modeling. 2010;**50**(2):240-250

[7] Antunes JE, Freitas MP, Rittner R. Bioactivities of a series of phosphodiesterase type 5 (PDE-5) inhibitors as modelled by MIA-QSAR. European Journal of Medicinal Chemistry. 2008;**43**(8):1632-1638

[8] Gaurav A, Singh R. 3D QSAR pharmacophore, CoMFA and CoMSIA based design and docking studies on phenyl alkyl ketones as inhibitors of phosphodiesterase 4. Medicinal chemistry (Shariqah (United Arab Emirates)). 2012;**8**(5):894-912

[9] Xing M, Akowuah GA, Gautam V, Gaurav A. Structure-based design of selective phosphodiesterase 4B inhibitors based on ginger phenolic compounds. Journal of Biomolecular Structure & Dynamics. 2016:1-15

[10] Hassaan EA, Sigler SC, Ibrahim TM, et al. Mining ZINC database to discover potential phosphodiesterase 9 inhibitors using Structure-Based drug design approach. Medicinal Chemistry (Shariqah (United Arab Emirates)). 2016;**12**(5):472-477

[11] Rauf A, Orhan IE, Ertas A, et al. Elucidation of Phosphodiesterase-1 inhibitory effect of some selected natural polyphenolics using in vitro and in silico methods. Current Topics in Medicinal Chemistry. 2017;**17**(4):412-417

[12] Nunes IK, de Souza ET, Cardozo SV, et al. Synthesis, Pharmacological profile and docking studies of new sulfonamides designed as Phosphodiesterase-4 inhibitors. PLoS One. 2016;**11**(10):e0162895

[13] Kumar J, Umar T, Kausar T, Mobashir M, Nayeem SM, Hoda N. Identification of lead BAY60-7550 analogues as potential inhibitors that utilize the hydrophobic groove in PDE2A: A molecular dynamics simulation study. Journal of Molecular Modeling. 2017;**23**(1):7

[14] Mondal C, Halder AK, Adhikari N, Jha T. Structural findings of cinnolines as anti-schizophrenic PDE10A inhibitors through comparative chemometric modeling. Molecular Diversity. 2014;**18**(3):655-671

[15] Heritage TW, Lowis DR. Molecular Hologram QSAR. Rational Drug Design. Vol 719. American Chemical Society; 1999. pp. 212-225

[16] Giembycz MA, Newton R. Harnessing the clinical efficacy of phosphodiesterase 4 inhibitors in inflammatory lung diseases: Dual-selective phosphodiesterase inhibitors and novel combination therapies. Phosphodiesterases as Drug Targets. Springer; 2011. pp. 415-446

[17] Giembycz MA, Field SK. Roflumilast: First phosphodiesterase 4 inhibitor approved for treatment of COPD. Drug Design, Development and Therapy. 2010;**4**:147

[18] Bamette MS, Bartus JOL, Burman M, et al. Association of the anti-inflammatory activity of phosphodiesterase 4 (PDE4) inhibitors with either inhibition of PDE4 catalytic activity or competition for [3 H] rolipram binding. Biochemical Pharmacology. 1996;**51**(7):949-956

[19] Naganuma K, Omura A, Maekawara N, et al. Discovery of selective PDE4B inhibitors. Bioorganic & Medicinal Chemistry Letters. 2009;**19**(12):3174-3176

[20] Jin S-LC, Goya S, Nakae S, et al. Phosphodiesterase 4B is essential for T H 2-cell function and development of airway hyperresponsiveness in allergic asthma. Journal of Allergy and Clinical Immunology. 2010;**126**(6):1252-1259. e1212

[21] Sharma V, Kumar H, Wakode S. Pharmacophore generation and atom based 3D-QSAR of quinoline derivatives as selective phosphodiesterase 4B inhibitors. Royal Society of Chemistry Advances. 2016;**6**(79):75805-75819

[22] Smith SJ, Brookes-Fazakerley S, Donnelly LE, Barnes PJ, Barnette MS, Giembycz MA. Ubiquitous expression of phosphodiesterase 7A in human proinflammatory and immune cells. American Journal of Physiology-Lung Cellular and Molecular Physiology. 2003;**284**(2):L279-L289

[23] Safavi M, Baeeri M, Abdollahi M. New methods for the discovery and synthesis of PDE7 inhibitors as new drugs for neurological and inflammatory disorders. Expert Opinion on Drug Discovery. 2013;**8**(6):733-751

[24] Dong H, Zitt C, Auriga C, Hatzelmann A, Epstein PM. Inhibition of PDE3, PDE4 and PDE7 potentiates glucocorticoid-induced apoptosis and overcomes glucocorticoid resistance in CEM T leukemic cells. Biochemical Pharmacology. 2010;**79**(3):321-329

[25] Pekkinen M, Ahlström ME, Riehle U, Huttunen MM, Lamberg-Allardt CJ. Effects of phosphodiesterase 7 inhibition by RNA interference on the gene expression and differentiation of human mesenchymal stem cell-derived osteoblasts. Bone. 2008;**43**(1):84-91

[26] Cichero E, Brullo C, Bruno O, Fossa P. Exhaustive 3D-QSAR analyses as a computational tool to explore the potency and selectivity profiles of thieno [3, 2-d] pyrimidin-4 (3 H)-one derivatives as PDE7 inhibitors. Royal Society of Chemistry Advances. 2016;**6**(66):61088-61108

[27] Cichero E, D'Ursi P, Moscatelli M, et al. Homology modeling, docking studies and molecular dynamic simulations using graphical processing unit architecture to probe the type-11 phosphodiesterase catalytic site: A computational approach for the rational design of selective inhibitors. Chemical Biology & Drug Design. 2013;**82**(6):718-731

[28] Maw GN, Allerton CM, Gbekor E, Million WA. Design, synthesis and biological activity of β-carboline-based type-5 phosphodiesterase inhibitors. Bioorganic & Medicinal Chemistry Letters. 2003;**13**(8):1425-1428

[29] Weeks JL, Corbin JD, Francis SH. Interactions between cyclic nucleotide phosphodiesterase 11 catalytic site and substrates or tadalafil and role of a critical Gln-869 hydrogen bond. Journal of Pharmacology and Experimental Therapeutics. 2009;**331**(1):133-141

[30] Ceyhan O, Birsoy K, Hoffman CS. Identification of biologically active PDE11-selective inhibitors using a yeast-based high-throughput screen. Chemistry & Biology. 2012;**19**(1): 155-163

[31] Moro S, Deflorian F, Bacilieri M, Spalluto G. Ligand-based homology modeling as attractive tool to inspect GPCR structural plasticity. Current Pharmaceutical Design. 2006;**12**(17):2175-2185

[32] Wang H, Liu Y, Huai Q, et al. Multiple Conformations of Phosphodiesterase-5 Implications for Enzyme Function and Drug Development. Journal of Biological Chemistry. 2006;**281**(30): 21469-21479

[33] Miller CL, Yan C. Targeting cyclic nucleotide phosphodiesterase in the heart: Therapeutic implications. Journal of Cardiovascular Translational Research. 2010;**3**(5):507-515

[34] Berman H.M, Westbrook J, Feng Z, Gilliland G, Bhat T.N, Weissig H, Shindyalov I.N, Bourne P.E. (2000) The Protein Data Bank Nucleic Acids Research, 28: 235-242. URL: www.rcsb.org Citation

[35] Li J, Zhou N, Liu W, et al. Discover natural compounds as potential phosphodiesterase-4B inhibitors via computational approaches. Journal of Biomolecular Structure & Dynamics. 2016;**34**(5):1101-1112

[36] Sarazan RD, Crumb WJ, Beasley CM, et al. Absence of clinically important HERG channel blockade by three compounds that inhibit phosphodiesterase 5—sildenafil, tadalafil, and vardenafil. European Journal of Pharmacology. 2004;**502**(3):163-167

[37] Witchel HJ. Drug-induced hERG Block and Long QT Syndrome. Cardiovascular Therapeutics. 2011;**29**(4):251-259

[38] Kayık G, Tüzün NS, Durdagi S. In silico design of novel hERG-neutral sildenafil-like PDE5 inhibitors. Journal of Biomolecular Structure and Dynamics. 2016:1-23

[39] Kayik G, Tuzun NS, Durdagi S. Investigation of PDE5/PDE6 and PDE5/PDE11 selective potent tadalafil-like PDE5 inhibitors using combination of molecular modeling approaches, molecular fingerprint-based virtual screening protocols and structure-based pharmacophore development. Journal of Enzyme Inhibition and Medicinal Chemistry. 2017;**32**(1):311-330

[40] Vemulapalli S, Watkins RW, Chintala M, et al. Antiplatelet and antiproliferative effects of SCH 51866, a novel Type 1 and Type 5 phosphodiesterase inhibitor. Journal of Cardiovascular Pharmacology. 1996;**28**(6):862-869

[41] Sybertz E, Czarniecki M. Inhibitors of PDE1 and PDE5 cGMP phosphodiesterases: Patents and therapeutic potential. Expert Opinion on Therapeutic Patents. 1997;**7**(6):631-639

[42] Yamazaki K, Kusunose N, Fujita K, et al. Identification of phosphodiesterase-1 and 5 dual inhibitors by a ligand-based virtual screening optimized for lead evolution. Bioorganic & Medicinal Chemistry Letters. 2006;**16**(5):1371-1379

[43] Dan A, Shiyama T, Yamazaki K, et al. Discovery of hydroxamic acid analogs as dual inhibitors of phosphodiesterase-1 and -5. Bioorganic & Medicinal Chemistry Letters. 2005;**18**:4085-4090

[44] Amin SA, Bhargava S, Adhikari N, Gayen S, Jha T. Exploring pyrazolo[3,4-d]pyrimidine phosphodiesterase 1 (PDE1) inhibitors: A predictive approach combining comparative validated multiple molecular modeling techniques. Journal of Biomolecular Structure & Dynamics. 2017:1-62

[45] Snyder GL, Prickaerts J, Wadenberg ML, et al. Preclinical profile of ITI-214, an inhibitor of phosphodiesterase 1, for enhancement of memory performance in rats. Psychopharmacology. 2016;**233**(17):3113-3124

[46] Card GL, England BP, Suzuki Y, et al. Structural basis for the activity of drugs that inhibit phosphodiesterases. Structure. 2004; **12**(12):2233-2247

[47] Manallack DT, Hughes RA, Thompson PE. The next generation of phosphodiesterase inhibitors: Structural clues to ligand and substrate selectivity of phosphodiesterases. Journal of Medicinal Chemistry. 2005;**48**(10):3449-3462

[48] Maurice DH, Ke H, Ahmad F, Wang Y, Chung J, Manganiello VC. Advances in targeting cyclic nucleotide phosphodiesterases. Nature Reviews Drug Discovery. 2014;**13**(4): 290-314

[49] Zhang KY, Card GL, Suzuki Y, et al. A glutamine switch mechanism for nucleotide selectivity by phosphodiesterases. Molecular Cell. 2004;**15**(2):279-286

[50] Li P, Zheng H, Zhao J, et al. Discovery of potent and selective inhibitors of phosphodiesterase 1 for the treatment of cognitive impairment associated with neurodegenerative and neuropsychiatric diseases. Journal of Medicinal Chemistry. 2016;**59**(3):1149-1164

[51] Xia Y, Chackalamannil S, Czarniecki M, et al. Synthesis and evaluation of polycyclic pyrazolo[3,4-d]pyrimidines as PDE1 and PDE5 cGMP phosphodiesterase inhibitors. Journal of Medicinal Chemistry. 1997;**40**(26):4372-4377

[52] Fisher DA, Smith JF, Pillar JS, St Denis SH, Cheng JB. Isolation and characterization of PDE9A, a novel human cGMP-specific phosphodiesterase. The Journal of Biological Chemistry. 1998;**273**(25):15559-15564

[53] Singh N, Patra S. Phosphodiesterase 9: Insights from protein structure and role in therapeutics. Life Sciences. 2014;**106**(1-2):1-11

[54] Wunder F, Tersteegen A, Rebmann A, Erb C, Fahrig T, Hendrix M. Characterization of the first potent and selective PDE9 inhibitor using a cGMP reporter cell line. Molecular Pharmacology. 2005;**68**(6):1775-1781

[55] Li Z, Lu X, Feng L-J, et al. Molecular dynamics-based discovery of novel phosphodiesterase-9A inhibitors with non-pyrazolopyrimidinone scaffolds. Molecular BioSystems. 2015;**11**(1):115-125

[56] Gomez L, Breitenbucher JG. PDE2 inhibition: Potential for the treatment of cognitive disorders. Bioorganic & Medicinal Chemistry Letters. 2013;**23**(24):6522-6527

[57] Yang B, Hamza A, Chen G, Wang Y, Zhan2 C-G. Computational determination of binding structures and free energies of phosphodiesterase-2 with Benzo[1,4]diazepin-2-one Derivatives. Journal of Physical Chemistry B. 2010;**114**(48):16020-16028

[58] Hamza A, Zhan C. Determination of the structure of human phosphodiesterase-2 in a bound state and its binding with inhibitors by molecular modeling, docking, and dynamics simulation. The Journal of Physical Chemistry B. 2009;**113**(9):2896-2908

[59] Naïm M, Bhat S, Rankin K, et al. Solvated interaction energy (SIE) for scoring protein-ligand binding affinities. 1. Exploring the parameter space. Journal of Chemical Information and Modeling 2007;**47**(1):122-133

[60] Podzuweit T, Nennstiel P, Muller A. Isozyme selective inhibition of cGMP-stimulated cyclic nucleotide phosphodiesterases by erythro-9-(2-hydroxy-3-nonyl) adenine. Cellular Signalling. 1995;**7**(7):733-738

[61] Rivet-Bastide M, Vandecasteele G, Hatem S, et al. cGMP-stimulated cyclic nucleotide phosphodiesterase regulates the basal calcium current in human atrial myocytes. Journal of Clinical Investigation. 1997;**99**(11):2710-2718

[62] Rutten K, Prickaerts J, Hendrix M, van der Staay FJ, Sik A, Blokland A. Time-dependent involvement of cAMP and cGMP in consolidation of object memory: Studies using selective phosphodiesterase type 2, 4 and 5 inhibitors. European Journal of Pharmacology. 2007;**558**(1-3):107-112

[63] Chambers RJ, Abrams K, Garceau NY, et al. A new chemical tool for exploring the physiological function of the PDE2 isozyme. Bioorganic & Medicinal Chemistry Letters. 2006;**16**(2):307-310

[64] Seybold J, Thomas D, Witzenrath M, et al. Tumor necrosis factor-alpha-dependent expression of phosphodiesterase 2: Role in endothelial hyperpermeability. Blood. 2005;**105**(9):3569-3576

[65] Buijnsters P, Angelis MD, Langlois X, et al. Structure-Based design of a potent, selective, and brain penetrating PDE2 inhibitor with demonstrated target engagement. ACS Medicinal Chemistry Letters. 2014;**5**(9):1049-1053

[66] Rombouts FJ, Tresadern G, Buijnsters P, et al. Pyrido[4,3-e][1,2,4]triazolo[4,3-a]pyrazines as selective, brain penetrant phosphodiesterase 2 (PDE2) inhibitors. ACS Medicinal Chemistry Letters. 2015;**6**(3):282-286

[67] Zhang C, Feng L-J, Huang Y, et al. Discovery of novel Phosphodiesterase-2A inhibitors by StructureBased virtual screening, structural optimization, and bioassay. Journal of Chemical Information and Modeling. 2017;**57**(2):355-364

[68] Bairoch A, Apweiler R. The SWISS-PROT protein sequence database and its supplement TrEMBL in 2000. Nucleic Acids Research. 2000;**28**(1):45-48

[69] Thompson JD, Higgins DG, Gibson TJ. CLUSTAL W: Improving the sensitivity of progressive multiple sequence alignment through sequence weighting, position-specific gap penalties and weight matrix choice. Nucleic Acids Research. 1994;**22**(22):4673-4680

[70] Morgenstern B, Frech K, Dress A, Werner T. DIALIGN: Finding local similarities by multiple sequence alignment. Bioinformatics (Oxford, England). 1998;**14**(3):290-294

[71] Jones DT. GenTHREADER: An efficient and reliable protein fold recognition method for genomic sequences. Journal of Molecular Biology. 1999;**287**(4):797-815

[72] Rost B, Sander C. Prediction of protein secondary structure at better than 70% accuracy. Journal of Molecular Biology. 1993;**232**(2):584-599

[73] Jones DT. Protein secondary structure prediction based on position-specific scoring matrices. Journal of Molecular Biology. 1999;**292**(2):195-202

[74] Sali A, Blundell TL. Comparative protein modelling by satisfaction of spatial restraints. Journal of Molecular Biology. 1993;**234**(3):779-815

[75] Laskowski RA, MacArthur MW, Moss DS, Thornton JM. PROCHECK: A program to check the stereochemical quality of protein structures. Journal of Applied Crystallography. 1993;**26**(2):283-291

[76] Kim KY, Lee H, Yoo SE, Kim SH, Kang NS. Discovery of new inhibitor for PDE3 by virtual screening. Bioorganic & Medicinal Chemistry Letters. 2011;**21**(6):1617-1620

Chapter 3

Discovery of BACE1 Inhibitors for the Treatment of Alzheimer's Disease

Yoshio Hamada and Yoshiaki Kiso

Additional information is available at the end of the chapter

http://dx.doi.org/10.5772/intechopen.68659

Abstract

Alzheimer's disease is the most common cause of dementia. According to the amyloid hypothesis, β-secretase (BACE1) is a promising molecular target for the development of anti-Alzheimer's disease drugs. BACE1 triggers the formation of the amyloid-β (Aβ) peptides that are the main component of senile plaques in the brain of patients with Alzheimer's disease. As BACE1 cleaves the amyloid precursor protein at the N-terminus of the Aβ domain, BACE1 inhibitors reduce the Aβ level in the brain. Previously, we designed a series of peptidic inhibitors that possessed a substrate transition-state analogue, and the structure-activity relationship of our inhibitors was evaluated, based on docking and scoring, using the docking simulation software Molecular Operating Environment (MOE). However, there was no association between the scoring values and the inhibitory activities at the P_2 position. Hence, we hypothesized that the interaction of the P_2 position of the inhibitor with the S_2 site of BACE1 was critical for the mechanism of inhibition, and we proposed the novel concept of 'electron donor bioisostere' for drug discovery. Using this concept, we designed potent small molecule non-peptidic BACE1 inhibitors.

Keywords: Alzheimer's disease, BACE1 inhibitor, docking simulation, electron donor bioisostere, *in-silico* conformational structure-based design

1. Introduction

Alzheimer's disease (AD), which is the most common cause of dementia, is characterized by progressive intellectual deterioration. In 1901, Alois Alzheimer, a psychiatrist and neuropathologist, observed a 51-year-old female patient at Frankfurt Asylum. The patient showed strange behavioural symptoms and the loss of short-term memory, which was later called 'AD'. Unfortunately, the cause of AD was unclear until recently, and there have been no treatment approaches since that first report by Dr. Alzheimer over 100 years ago. Recently, the development of many drug candidates based on the amyloid hypothesis has been reported. β-secretase (BACE1; β-site amyloid precursor protein (APP) cleaving enzyme 1) is a promising molecular target for the development of anti-Alzheimer's drugs. BACE1 triggers the formation of the amyloid-β (Aβ) peptide that is the main component of the senile plaques found in the brain of AD patients. Previously, we had designed a series of peptidic inhibitors that possessed a substrate transition-state analogue, and evaluated the structure-activity relationship of our inhibitors, based on docking and scoring, using the docking simulation software Molecular Operating Environment ((MOE), Chemical Computing Group Inc., Canada).

1.1. Pathology of Alzheimer's disease

Although the cause of AD was unclear until recently, a breakthrough was obtained from the genetic study of some patients with familial AD. Certain mutations of the amyloid precursor protein (APP) or presenilin gene increased amyloid-β peptides (Aβs) in the brain, which indicated their involvement in the pathogenesis of AD [1–4]. Aβ is produced from APP by two processing enzymes, β-secretase and γ-secretase, which are potential molecular targets for the development of anti-AD drugs [5–7]. The cleavage sites of APP are shown in **Figure 1A**. BACE1, one of the processing enzymes of APP, triggers Aβ formation in the rate-limiting first step by the cleavage of APP at the Aβ domain N-terminus (β-site). BACE1 is a type-I transmembrane

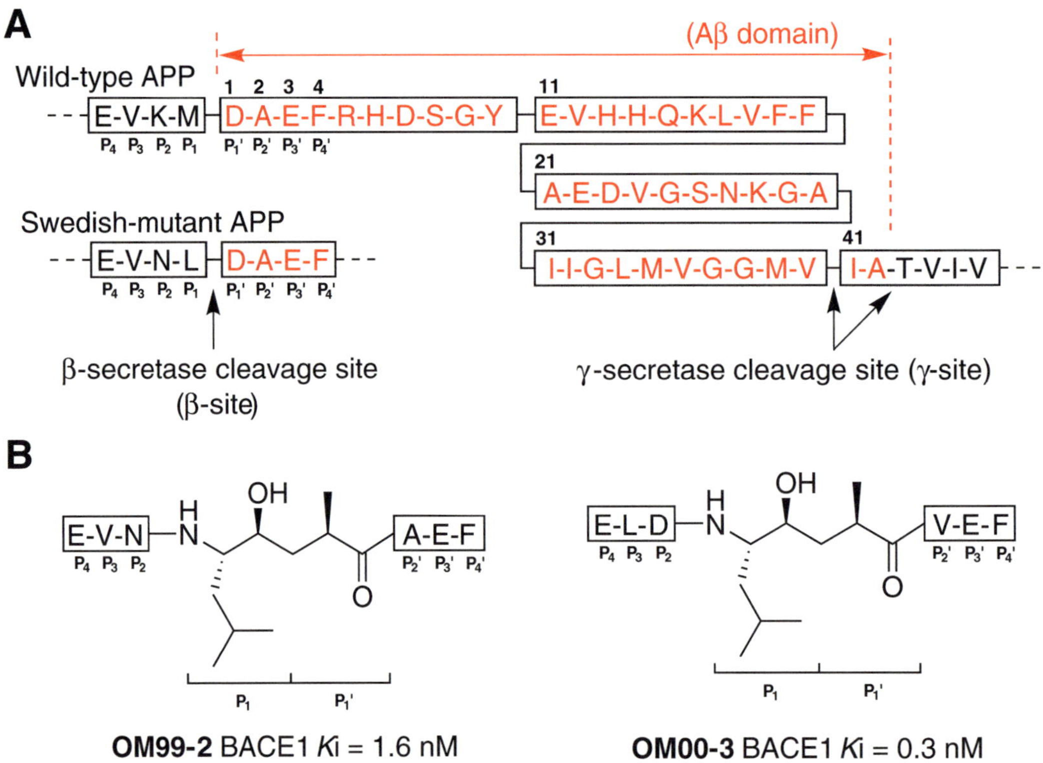

Figure 1. (A) Amyloid precursor protein (APP) and its cleavage site. (B) Early peptidic BACE1 inhibitor by Ghosh et al.

aspartic protease with 501 amino acids. BACE1 and APP are located in the same intracellular granules, the endoplasmic reticulum, Golgi, and trans-Golgi networks, which are acidic environments, which suggested that Aβs are produced in these locations [8]. Next, another aspartic protease, the γ-secretase complex, cleaved at the C-terminus of the Aβ domain and released Aβ peptides. The γ-secretase complex that contains a protein *via* the presenilin gene as a catalytic component cleaved at two cleavage sites, 'γ-sites', which mainly resulted in the formation of two species of Aβs: $A\beta_{1\text{-}40}$ and $A\beta_{1\text{-}42}$ (**Figure 1A**). $A\beta_{1\text{-}42}$ shows greater neurotoxicity and aggregability than $A\beta_{1\text{-}40}$ and appears to be a key biomolecular marker of AD pathogenesis. According to the amyloid hypothesis, BACE1 and γ-secretase appear to be molecular targets for the development of anti-AD drugs. However, because γ-secretase can cleave other single-pass transmembrane proteins *in vivo*, such as Notch, which plays a critical role in cell differentiation, γ-secretase inhibitors appeared to lead to serious side effects. As BACE1 knockout transgenic mice demonstrated normal survival, this indicated a promising direction of study, in which BACE1 is a molecular target for the development of AD drugs [9]. At present, many BACE1 inhibitors have been revealed, including those in our study [10–16].

1.2. Early peptidic BACE1 inhibitors

An early inhibitor of BACE1, an aspartic protease, was designed on the basis of a substrate transition-state concept, as well as that of other aspartic proteases, such as renin and HIV protease, which have a substrate transition-state analogue at the P_1 position [10–16]. It is well-known that the Swedish mutant APP (K670N and M671L double mutation, **Figure 1A**) is cleaved faster than wild-type APP by BACE1, which results in increased $A\beta_{1\text{-}42}$ and $A\beta_{1\text{-}40}$ levels. Early BACE1 inhibitors were designed based on the Swedish-mutant APP amino acid sequence. In 1999, Sinha et al. from Elan Pharmaceuticals purified the BACE1 enzyme from the human brain using a transition-state analogue based on the Swedish-mutant sequence, and succeeded in cloning the BACE1 enzyme [17]. Ghosh et al. reported the potent inhibitors **OM99-2** (*K*i = 1.6 nM) and **OM00-3** (*K*i = 0.3 nM) with a hydroxyethylene unit as a substrate transition-state analogue (**Figure 1B**) and the first X-ray crystal structure (PDB ID: 1FKN) of a complex between recombinant BACE1 and OM99-2 [18–21].

We have reported a series of peptidic BACE1 inhibitors that possessed a norstatine-type transition-state analogue [22–30]. Our early inhibitors are shown in **Table 1**. Octapeptide **1** with an Asn residue and (2*R*, 3*S*)-3 amino-2-hydroxy-5-methylhexanoic acid (Nst, Leu-type transition-state analogue) at the P_2 and P_1 positions, respectively, corresponding to the Swedish-mutant APP sequence showed no inhibitory activity. Octapeptide **2** with (2*R*, 3*S*)-3-amino-2-hydroxy-4-phenylbutyric acid (Pns) as a transition-state analogue at the P_1 position, showed weak inhibitory activity. Compound **3** with an Asp residue similar to OM00-3, and compound **4**, with a Met residue at the P_2 position, also showed weak inhibitory activity. Although compound **5** with the P_2-Lys residue that corresponded to the wild-type APP sequence showed no inhibitory activity, octapeptide **6** that possessed a Leu residue at the P_2 position exhibited potent inhibitory activity (>90% at 2 μM). We synthesized some truncated peptides on the N- or C-terminus in order to confirm the essential moiety for the inhibitory effect. N-truncation of peptides eliminated their inhibitory activity (peptides **7-8**). Although C-truncated peptides **10-13** showed a weaker inhibitory activity than octapeptide **6**, pentapeptide **12** replicated the inhibitory activity

Compound	P_4	P_3	P_2	P_1	P_1'	P_2'	P_3'	P_4'	BACE1 inhibition (%)[1]
1	E	V	N[2]	Nst	D	A	E	F	<20
2	E	V	N[2]	Pns	D	A	E	F	24
3	E	V	D	Pns	D	A	E	F	25
4	E	V	M	Pns	D	A	E	F	42
5	E	V	K[3]	Pns	D	A	E	F	<20
6	E	V	L	Pns	D	A	E	F	>90
7		V	L	Pns	D	A	E	F	<20
8			L	Pns	D	A	E	F	<20
9				Pns	D	A	E	F	<20
10	E	V	L	Pns	D	A	E		60
11	E	V	L	Pns	D	A			46
12	E	V	L	Pns	D				61
13	E	V	L	Pns					34

[1]BACE1 inhibition activities at 2 μM.

[2]P_2 residue corresponding to the Swedish-mutant APP sequence.

[3]P_2 residue corresponding to the wild-type APP sequence.

Table 1. BACE1 inhibitory activities of peptidic BACE1 inhibitors.

of heptapeptide **10**. Hence, we designed a series of pentapeptidic BACE1 inhibitors based on compound **12** using a computational approach.

2. Docking simulation and design of pentapeptidic BACE1 inhibitors

Early BACE1 inhibitors were designed using the coordinates of the first reported X-ray crystal structure (1FKN) of a complex between BACE1 and OM99-2. OM99-2 has an Asn residue, which corresponded to the P_2 residue of Swedish mutant sequence. As OM99-2 in 1FKN appeared to interact with the BACE1-Arg235 side chain *via* hydrogen bonding, many researchers have designed BACE1 inhibitors that possessed a hydrogen bond receptor, such as an Asn residue, at the P_2 position, using the 1FKN coordinates. However, our peptidic BACE1 inhibitors with an Asn at the P_2 position showed no inhibitory activity, and, as shown in **Table 1**, the P_2 residue that showed potent BACE1 inhibitory activity was a hydrophobic amino acid residue, Leu. Thus, our design strategy seemed to require a fundamental review. We researched the inhibitory mechanism of our peptides using a computational approach. As we found that pentapeptide **21**, which possessed an aminobenzoic acid residue as a bio-isostere of the Asp residue at the P_1' position, showed higher inhibitory activity than penta-peptide **12**, we evaluated the series of pentapeptides **14-24** with an aminobenzoic acid residue by using a docking simulation, and then synthesized the compounds (**Table 2**). The docking simulation was performed using MOE software under the MMFF94x force field. The calculated active sites of BACE1 were depicted in **Figure 2A** using the Alpha SiteFinder application in MOE software and the coordinate set of X-ray crystal structure, 1FKN. The 3D structure of OM99-2 after a docking simulation is shown in **Figure 2B** as a magenta-coloured stick model. Because the moieties from the P_1 to P_4 positions of OM99-2 almost coincided with that of the X-ray crystal structure (aqua coloured stick model) of OM99-2, we performed the docking simulation study using this calculation model. Although the moieties from the P_2' to P_4' positions of OM99-2 assumed a different pose from that of the X-ray crystal structure of OM99-2, their moieties were placed outside the BACE1 enzyme. It is likely that the difference between the X-ray crystal structure and the docking calculation might occur by a packing at the crystallization of BACE1 complex. The results of the docking simulation are shown in **Table 2**. Peptides **14-24** and OM99-2 were scored using the scoring function in the MOE software.

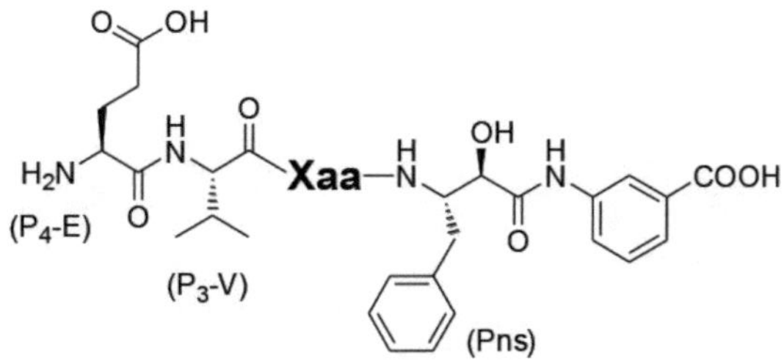

Compound	Xaa	U_ele[1]	U_vdw[2]	U_str[3]	U_dock[4]	BACE1 inhibition (%)[5]
OM99-2 (see **Figure 1**)		−284.7	28.0	14.6	−242.1	—
14	N	−184.4	4.2	10.2	−170.0	28.4
15	M	−196.1	14.8	11.4	−169.9	63.9
16	Y	−195.2	18.5	13.6	−163.1	36.9
17	D	−174.1	2.0	11.2	−160.9	33.2
18	I	−196.3	25.7	11.1	−159.5	65.8
19	F	−176.8	9.0	11.6	−156.2	47.3
20	E	−179.8	11.9	22.3	−145.6	36.3
21	L	−155.1	3.3	10.2	−141.6	83.7
22	W	−187.1	19.6	26.6	−140.9	71.3
23	Q	−148.9	−0.3	9.8	−139.4	14.1
24	Cha[6]	−147.1	−1.2	9.7	−138.6	84.1

[1]Electrostatic energy between BACE1 and inhibitor (kcal/mol).
[2]van der Waals energy between BACE1 and inhibitor (kcal/mol).
[3]Strain energy of inhibitor (kcal/mol).
[4]Docking score (kcal/mol); U_dock = U_ele + U_vdw + U_str.
[5]BACE1 inhibition % at 2 μM.
[6]Cyclohexylalanine (Cha).

Table 2. Docking simulation of pentapeptidic BACE1 inhibitors and their score using the 1FKN X-ray crystal structure.

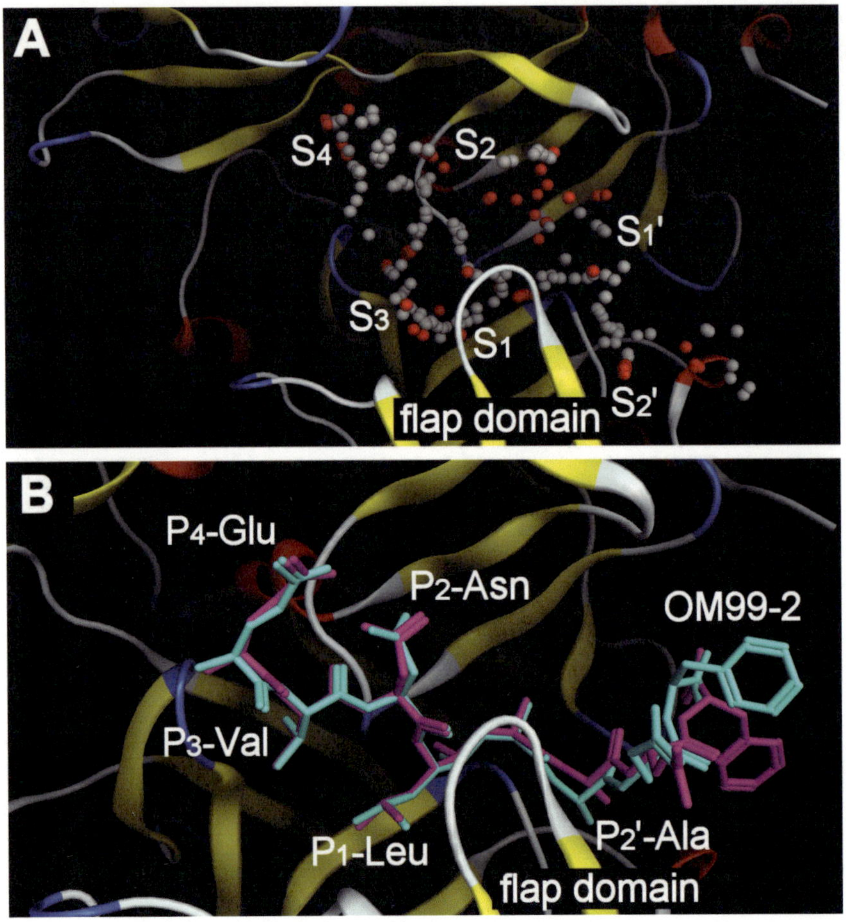

Figure 2. (A) Calculated active sites using the Alpha Site Finder application in MOE software. (B) OM99-2 docked in BACE1. Aqua and magenta colours indicate the X-ray crystal structure 1FKN and the energy-minimized structure under the MMFF94x force field, respectively.

U_ele, U_vdw, and U_str indicated the potential energies of electrostatic interaction, van der Waals interactions between enzyme and ligand, and strain energy of the ligand, respectively, and their sum is a docking score, U_dock. OM99-2 and peptide **14**, which possess an Asn residue at the P_2 position, showed good scores. However, peptides **14** and **17**, that possess a hydrophilic amino acid, such as Asn and Asp, showed a low inhibitory activity. The residues of peptides **21** and **24** that exhibited a high inhibitory activity were hydrophobic amino acids such as Leu and cyclohexylalanine (Cha), as well as the results in **Table 1**. Interestingly, peptides **21** and **24** showed low scores. Thus, there was no association between the scoring values and the inhibitory activity at the P_2 position.

As the docking model using the coordinate set of 1FKN appeared to give an unfavourable score for the BACE1 inhibitor, we compared the publicly available X-ray crystal structures of BACE1-inhibitor complexes. Surprisingly, the guanidine group of BACE1-Arg235 in most crystal structures, except 1FKN, showed similar figures flopping over the P_2 region of the inhibitors, and the nearest distances between the guanidino-plane of Arg235 side chain and the P_2 region of the inhibitor showed similar values of approximately 3Å [31]. The P_2 moieties in most of the crystal structures found to interact with the BACE1-Arg235 side chain were a methyl group, carbonyl oxygen atom, or aromatic ring, which were bound to the guanidine-plane of Arg235 side chain by CH-π, O-π, or π-π stacking interactions. This suggested that the π-orbital on the guanidino-plane interacted with the P_2 region by a weak quantum force such as stacking or σ-π interaction. The only exception was the interaction in the first reported X-ray crystal structure, 1FKN. The P_2 moiety of OM99-2 in the crystal structure 1FKN appeared to interact with the BACE1-Arg235 side chain *via* hydrogen bonding (**Figure 3A**). OM00-3, which was reported by the same researchers, was an inhibitor that was structurally similar to OM99-2; surprisingly, the P_2-Asp side chain of OM00-3 docked in BACE1 (PDB ID: 1M4H) interacted with the π-orbital on the guanidine-plane of the BACE1-Arg235 side chain *via* O-π interaction (**Figure 3B**). Many early BACE1 inhibitors that possess a hydrogen bond receptor at the P_2 position were designed using the first reported crystal structure 1FKN. However, the hydrogen bonding interaction between most of the inhibitors and the BACE1-Arg235 side chain was not shown in their crystal structures. For instance, the inhibitor from Merck (MSD), crystal structure (PDB ID: 2B8L), interacted with the BACE1-Arg235 side chain *via* a CH-π interaction (**Figure 3C**). The researchers at MSD most likely based their inhibitor on a structure that possessed an *N*-methyl-*N*-methanesulfonyl group at the P_2 position in anticipation of the hydrogen-bonding interaction between the sulfonyl oxygen atom and the BACE1-Arg235 side chain. However, the *N*-methyl group of the MSD inhibitor interacted with the π-orbital on the guanidine-plane of the BACE1-Arg235 side chain at a distance of 2.8 Å. The inhibitor reported by Pfizer (PDB ID: 2P83) appeared to interact with the BACE1-Arg235 side chain *via* O-π interactions, as shown in **Figure 3D**. As seen above, most of the BACE1 inhibitors, except OM99-2 in the crystal structure 1FKN, interacted with the BACE1-Arg235 side chain by a weak quantum force such as stacking or σ-π interaction. The Arg235 side chain of the BACE1-OM99-2 complex (1FKN) assumed an exceptionally different pose to the other crystal structures because the BACE1 complex appears to be stabilized by intramolecular hydrogen-bonding interaction between the P_4-Glu and P_2-Asn side chains of OM99-2 (**Figure 3A**). Because OM00-3 does not form such intramolecular hydrogen-bonding, the P_2 residue of OM00-3 appears to interact with the BACE1-Arg235 side chain by a quantum chemical interaction. As many researchers have designed BACE1 inhibitors with a hydrogen bond receptor on the basis of the first reported crystal structure 1FKN, docking models using 1FKN will require further review. Furthermore, we found that the side chain of BACE1-Arg235 could move in concert with the inhibitor's size. The superimposed figure of four crystal structures (PDB ID: 2B8L, 1M4H, 1W51, and 2IQG) of the complex between BACE1 and the inhibitors is depicted in **Figure 4**. The guanidino-planes of BACE1-Arg235 in the crystal structures of most BACE1 complexes showed similar distances from the P_2 regions of the inhibitors regardless of their molecular size [31]. This fact suggested a serious issue for a docking simulation for the drug discovery of BACE1 inhibitors. However, the BACE1-Arg235 side chain seems to have a restricted range of motion: the BACE1-Arg235 side chain slides sideways, not up and down, along the wall of the β-sheet structure that consists of four peptide strands behind the flap domain of BACE1; therefore, the location of the BACE1-Arg235 side chain could be predicted by the inhibitor's size. As shown in **Figure 4**, the

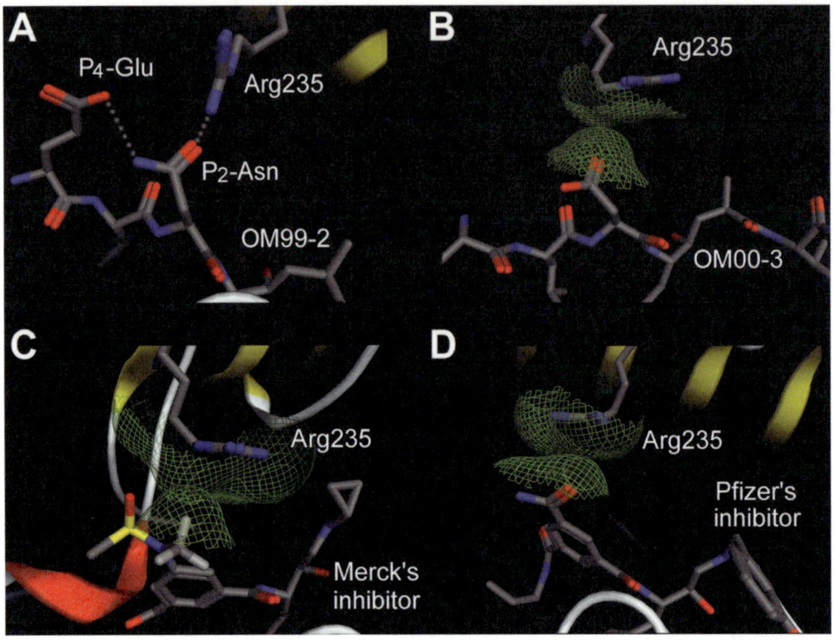

Figure 3. Interaction of BACE1-Arg235 with BACE1 inhibitors in X-ray crystal structures. (A) PDB ID: 1FKN, (B) PDB ID: 1M4H, (C) PDB ID: 2B8L and (D) PDB ID: 2P83.

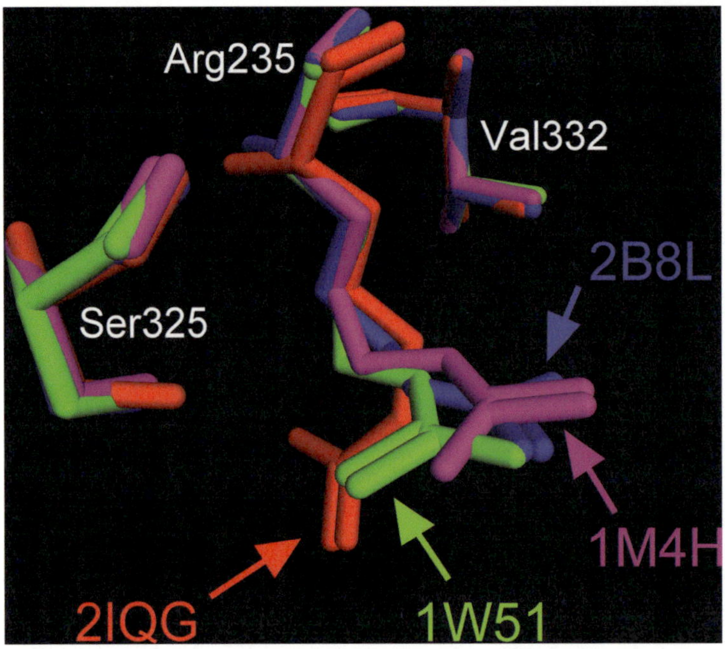

Figure 4. Location of BACE1-Arg235 in the respective crystal structures. The blue, magenta, green and red stick models represent the X-ray crystal structures of the BACE1-inhibitor complexes, 2B8L, 1M4H, 1W51 and 2IQG, respectively.

orbital follows the same pattern along the wall of the β-sheet structure of BACE1 and might be predicted according to the inhibitor's size. We hypothesized that the role of the BACE1-Arg235 side chain was important for the BACE1 inhibitory mechanism. The guanidine-plane of Arg235 that can move in concert with the inhibitor's size appears to push down on the P_2 region of the inhibitor, which caused them to be affixed to the active site of BACE1 because of this 'flop-over' mechanism by the BACE1-Arg235 side chain. Although a quantum chemical force, such as σ-π interaction, has a weaker binding energy than a hydrogen bonding interaction, this 'flop-over' mechanism permits a strong binding mode with the active site of BACE1.

For the reasons mentioned above, we performed a docking calculation using the X-ray crystal structure 1M4H, in which the P_2 moiety of the inhibitor (OM00-3) had a similar size to that of our inhibitor (**Table 3**). OM00-3 appears to show a high docking score value owing to its large molecular size: OM00-3 has many more amide bonds than our peptapeptidic inhibitors, and

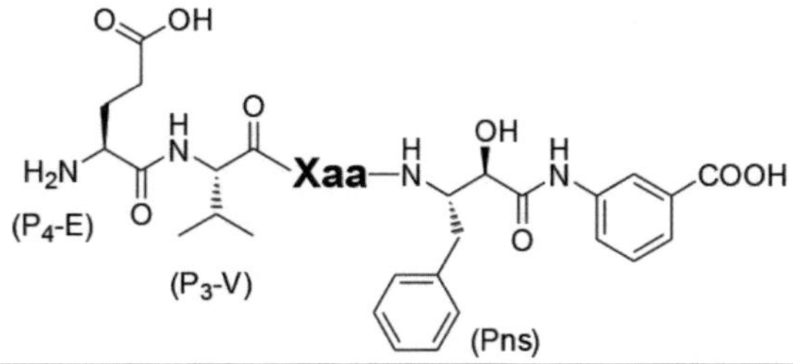

Compound	Xaa	U_ele[1]	U_vdw[2]	U_str[3]	U_dock[4]	BACE1 inhibition (%)[5]
OM00-3 (see Figure 1)		−233.4	0.7	15.7	−217.0	–
21	L	−195.7	11.3	10.3	−174.1	83.7
22	W	−189.3	13.0	10.5	−165.8	71.3
18	I	−194.6	18.5	11.1	−165.0	65.8
19	F	−195.6	23.0	11.7	−160.9	47.3
15	M	−195.6	24.6	10.1	−160.9	63.9
24	Cha[6]	−194.4	26.6	11.6	−156.2	84.1
16	Y	−196.7	26.5	14.1	−156.1	36.9
20	E	−190.7	24.2	10.6	−155.9	36.3
14	N	−190.7	24.2	11.3	−155.2	28.4
23	Q	−196.9	27.6	15.7	−153.6	14.1
17	D	−185.5	28.5	9.7	−147.3	33.2

[1]Electrostatic energy between BACE1 and inhibitor (kcal/mol).
[2]van der Waals energy between BACE1 and inhibitor (kcal/mol).
[3]Strain energy of inhibitor (kcal/mol).
[4]Docking score (kcal/mol); U_dock = U_ele + U_vdw + U_str.
[5]BACE1 inhibition % at 2 μM.
[6]Cyclohexylalanine (Cha).

Table 3. Docking simulation of pentapeptidic BACE1 inhibitors and their scoring using the 1M4H X-ray crystal structure.

can closely interact with the BACE1 active site *via* electrostatic energy. The correlation chart between BACE1 inhibitory activities and the docking score values is shown in **Figure 5**, and it indicated a good correlation coefficient ($r = -0.717$). Peptides that possess a hydrophilic amino acid residue at the P_2 position showed low docking score values, which indicated that these P_2 residues cannot interact with the BACE1-Arg235 side chain *via* electrostatic energy in the X-ray crystal structure 1M4H, and the docking score showed a good correlation with BACE1 inhibitory activity as a result. However, the plot of peptide **24** was outside of the correlation line. Peptide **24**, with the bulky amino acid Cha at the P_2 position, might show van der Waals repulsion against the BACE1-Arg235 side chain.

Furthermore, we designed a series of BACE1 inhibitors that possessed one or more bioisosteres of carboxylic acid from pentapeptide **21** that conferred excellent values to both docking score and BACE1 inhibitory activity, in order to develop practical BACE1 inhibitors as drug candidates (**Figure 6**). A tetrazole ring is known as a carboxylic bioisostere. Because it is well known that 5-aminotetrazole was highly explosive, peptides **25** and **26**, which possessed a carboxylic acid bioisostere at the P_4 position were designed and synthesized using tetrazole-5-carboxylic acid. Peptides **25** and **26** showed potent BACE1 inhibitory activities (IC_{50} = 8.2 nM and 3.9 nM, respectively) [24, 25]. Moreover, peptide **27**, which possessed two tetrazole rings on the P_1' ring, was synthesized. Peptide **27** showed the most potent BACE1 inhibitory activity (IC_{50} = 1.2 nM) [26].

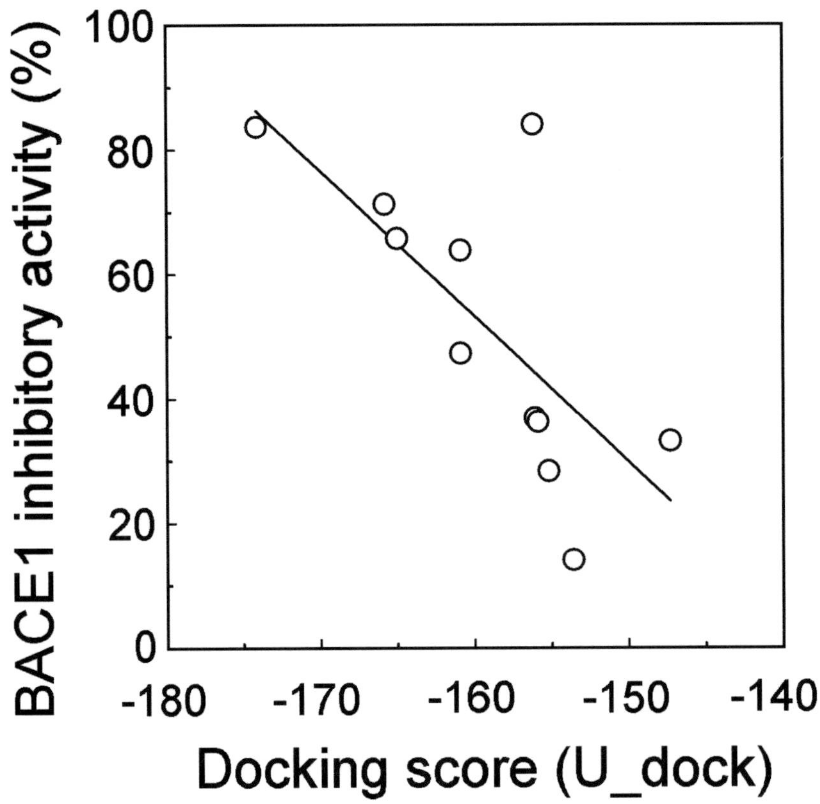

Figure 5. The correlation between BACE1 inhibitory activities and the docking score values.

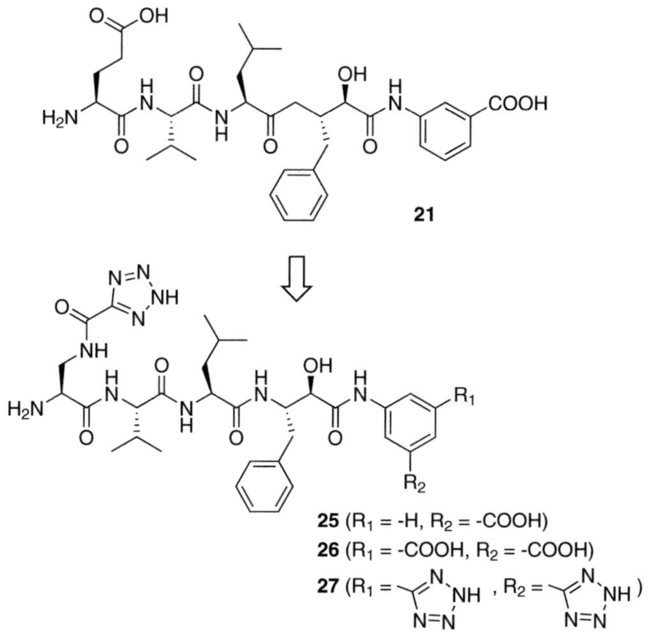

Compound	BACE1 inhibition (%)		IC_{50} (nM)
	at 2 μM	at 0.2 μM	
21	83.7	—	330
25	99.1	87.1	8.2
26	100	98.1	3.9
27	100	100	1.2

Figure 6. Design of practical BACE1 inhibitors using a carboxylic acid bioisostere.

3. Design of small-sized non-peptidic BACE1 inhibitors

At present, many non-peptidic BACE1 inhibitors have been discovered. The research of Elan pharmaceuticals, Merck (MSD), Pfizer, and Schering-Plough, and Ghosh et al. reported the BACE1 inhibitors **28-32** (IC_{50} = 15, 15, 5, 3, and 1 nM, respectively) with an isophthalic scaffold at the P_2 position, as shown in **Figure 7** [14, 16]. Because the distance between the flap domain and the cleft domain that forms the S_2 pocket of BACE1 was narrow, a planar aromatic ring, such as an isophthalic scaffold, might dock closely in the S_2 pocket of BACE1. Hence, we designed a series of BACE1 inhibitors from the virtual inhibitor **28** (**Figure 8**), in which the P_2 moiety of our peptidic inhibitors was replaced with an isophthalic scaffold [31–35]. First, we focused on the sterically hindered interaction between the P_3 amide and a proton on the P_2-isophthalic ring of the virtual inhibitor, which restricts the configuration. Using the approach '*in-silico* conformational structure-based design' based on a conformer of the docked inhibitor in BACE1, we adopted a pyridinedicarboxylic scaffold as a P_2 moiety, which lacked the 2-proton from the isophthalic ring, and designed inhibitor **29** [32]. Next, we optimized the P_3-region of inhibitor **29**

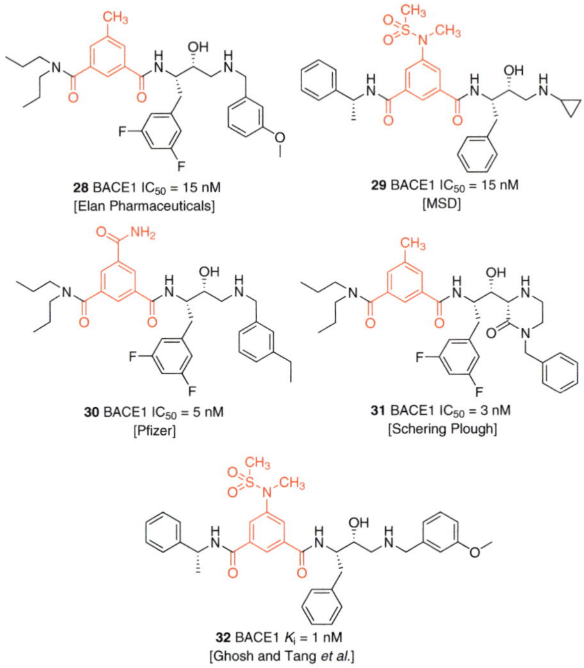

Figure 7. BACE1 inhibitors with an isophthalic scaffold.

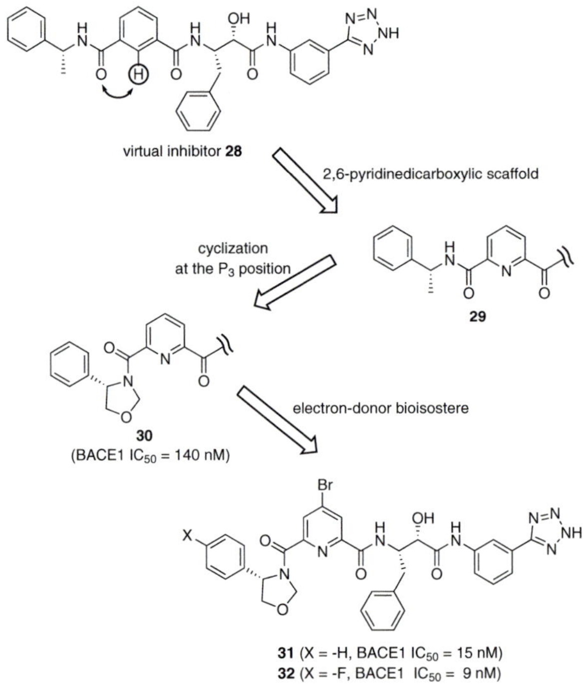

Figure 8. Design of small-sized non-peptidic BACE1 inhibitors with a 2,6-pyridinedicarboxylic scaffold.

with a P_2-pyridinedicarboxylic scaffold. There is a S_3 sub-pocket behind the active site of BACE1, and the P_3 phenyl group of **29** appears to interact with the S_3 sub-pocket. We envisioned that inhibitors with a P_3 benzylamide group assumed a folding pose between the P_2 aromatic scaffold and the P_3 benzylamide, and that the α-methyl group on the P_3 benzylamide of **29** stabilized this folding structure. Hence, we designed inhibitor **30** by the introduction of a five-membered ring, oxazolidine, at the P_3 region to fix the folding structure [32]. The oxazolidine ring fixes the direction of the phenyl ring at the P_3 position, so that the P_3 phenyl ring may be able to bind closely to the S_3 sub-pocket of BACE1. Inhibitor **30** showed moderate BACE1 inhibitory activity (IC_{50} = 140 nM).

Although *in-silico* approaches, such as a docking simulation between an enzyme/receptor and drugs, have contributed greatly to drug discovery research, most docking simulation software programs adopt molecular mechanics/molecular dynamics (MM/MD) calculations based on classical Newtonian mechanics. However, docking simulations using these calculations do not appear to estimate weak quantum chemical interactions, such as stacking or σ-π interaction, between the BACE1-Arg235 side chain and inhibitors. Because the Arg residue is recognized as charged in these software programs, the quantum chemical interactions involving an Arg side chain are unlikely to receive a favourable score. The concept of 'bioisostere' is important for the development of practical drugs. However, in the case of BACE1 inhibitor design, the bioisostere of the P_2 moiety, according to the Swedish-mutant APP, is an Asn or an amide residue based on a classical bioisostere concept that does not assume quantum chemical interactions. Therefore, inhibitors that can interact with the Arg235 side chain on the basis of a quantum chemical interaction could never be designed using such a classical concept. The researchers at Bristol-Myers Squibb (BMS) reported a series of BACE1 inhibitors that can interact with the BACE1-Arg235 side chain by π-π stacking. According to their SAR study, the introduction of an electron-donating methoxy group to the *p*-position of the phenyl ring that interacts with the BACE1-Arg235 side chain enhanced BACE1 inhibitory activity. This indicated that an inhibitor that possessed a P_2 aromatic ring with higher electron density could strongly dock to the active site of BACE1 that has an electron-poor π-orbital on the guanidino-plane of the BACE1-Arg235 side chain. Hence, we proposed a new concept of the 'electron-donor bioisostere', which can interact with an electron-poor π-orbital, such as the guanidine group of Arg235, by quantum chemical interactions [14].

Based on the electron-donor bioisostere concept, we speculated that an electron-rich halogen atom could interact with the electron-poor guanidine π-orbital by Coulomb interaction. Using the *ab initio* molecular orbital approach, Imai et al. indicated that the calculated Cl-π interaction energy was slightly stronger than that of CH-π interaction, and its energy was affected by π-electron density [36]. Inhibitor **31**, which possessed a halogen atom on the P_2 aromatic scaffold, exhibited potent inhibitory activity (IC_{50} = 15 nM). Moreover, inhibitor **32**, which possessed a fluorine atom on the *p*-position of the P_3 phenyl group exhibited the most potent inhibitory activity (IC_{50} = 9 nM), and was available from Wako Pure Chemical Industries (Japan) as a reagent for biological research [31]. The drastic improvement in the inhibition of BACE1 following the introduction of a halogen atom into the P_2 position of our compounds appears to support our hypothesis; namely, the quantum chemical interactions between BACE1 and its inhibitors play a critical role in the mechanism of BACE1 inhibition.

4. Conclusion

Although we calculated the docking scoring values by a docking calculation between BACE1 and its inhibitors using the first reported X-ray crystal structure 1FKN, we found no association between the scoring values and BACE1 inhibition. We found that a specific interaction, a quantum chemical interaction between the Arg235 side chain and the P_2 region of the inhibitor, played a critical role in the inhibitory mechanism of BACE1. Whereas most BACE1 inhibitors, except OM99-2, interacted with BACE1-Arg235 by a quantum chemical interaction, such as stacking and σ-π interaction, many early BACE1 inhibitors were designed using the 1FKN coordinate set. As the crystal structure 1FKN showed a hydrogen bonding between the BACE1-Arg235 side chain and OM99-2, the early studies on BACE1 inhibitor design might have misdirected, as a docking simulation using 1FKN appears to be meaningless. In fact, unlike 1FKN, there is no hydrogen bonding interaction present in most of the X-ray crystal structures. We selected the peptide sequence that showed potent inhibitory activity by a docking simulation using the X-ray crystal structure 1M4H, and designed potent peptidic BACE1 inhibitors with one or more carboxylic acid bioisosteres. Moreover, we focused on a quantum chemical interaction, and designed the potent non-peptidic BACE1 inhibitor **32** using the 'electron-donor bioisostere concept' that we have proposed. Our findings indicated the importance of the X-ray crystal structure in computational drug design.

Acknowledgements

This study was supported in part by the Grants-in-Aid for Scientific Research from MEXT (Ministry of Education, Culture, Sports, Science and Technology), Japan (KAKENHI No. 23590137 and No. 26460163), and a donation from Professor Emeritus Tetsuro Fujita of Kyoto University. At the time of writing, we received word that Prof. Fujita had passed away on January 1, 2017. Prof. Fujita was the teacher of one of the authors, Y. Hamada, and was known as the inventor of a treatment agent for multiple sclerosis. We dedicate this article to Prof. Fujita.

Author details

Yoshio Hamada[1*] and Yoshiaki Kiso[2]

*Address all correspondence to: pynden@gmail.com

1 Faculty of Frontier of Innovative Research in Science and Technology, Konan University, Japan

2 Laboratry of Peptide Science, Nagahama Institute of Bio-Science and Technology, Japan

References

[1] Selkoe DJ. Toward a comprehensive theory for Alzheimer's disease. Hypothesis: Alzheimer's disease is caused by the cerebral accumulation and cytotoxicity of amyloid β-protein. Annals of the New York Academy of Sciences. 2000;**924**:17-25. DOI: 10.1111/j.1749-6632.2000.tb05554.x

[2] Selkoe DJ. The deposition of amyloid proteins in the aging mammalian brain: Implications for Alzheimer's disease. Annals of Medicine. 1989;**21**:73-76. DOI: 10.3109/07853898909149187

[3] Selkoe DJ. Translating cell biology into therapeutic advances in Alzheimer's disease. Nature. 1999;**399**:A23-A31. DOI: 10.1038/399a023

[4] Sinha S, Lieberburg I. Cellular mechanisms of β-amyloid production and secretion. Proceedings of the National Academy of Sciences of the United States of America. 1999;**96**:11049-11053. DOI: 10.1073/pnas.96.20.11049

[5] Vassar R, Bennett BD, Babu-Khan S, et al. β-secretase cleavage of Alzheimer's amyloid precursor protein by the transmembrane aspartic protease BACE. Science. 1999:**286**;735-741. DOI: 10.1523/JNEUROSCI.3657-09.2009

[6] Hussain I, Powell D, Howlett DR, et al. Identification of a novel aspartic protease (Asp 2) as β-secretase. Neuroscience. 1999:**14**;419-427. DOI: 10.1523/JNEUROSCI.3657-09.2009

[7] Yan R, Bienkowski MJ, Shuck ME, et al. Membrane-anchored aspartyl protease with Alzheimer's disease β-secretase activity. Nature. 1999:**402**;533-537. DOI: 10.1038/990107

[8] Turner PR, O'Connor K, Tate WP, Abraham WC. Roles of amyloid precursor protein and its fragment in regulating neural activity, plasticity and memory. Progress in Neurobiology. 2003:**70**;1-32. DOI: 10.1016/S0301-0082(03)00089-3

[9] Roberds SL, Anderson J, Basi G, et al. BACE knockout mice are healthy despite lacking the primary β-secretase activity in brain: Implications for Alzheimer's disease therapeutics. Human Molecular Genetics. 2001:**10**;1317-1324. DOI: 10.1093/hmg/10.12.1317

[10] Hamada Y, Kiso Y. New directions for protease inhibitors directed drug discovery. Biopolymers. 2016:**106**;563-579. DOI: 10.1002/bip.22780

[11] Hamada Y, Kiso Y. Aspartic protease inhibitors as drug candidates for treating various difficult-to-treat diseases. In: Ryadnov M, Farkas E, editors. Amino Acids, Peptides and Proteins. Vol. 39. London: Royal Society of Chemistry; 2015. pp. 114-147. DOI: 10.1039/9781849739962-00114

[12] Hamada Y. Drug discovery of b-secretase inhibitors based on quantum chemical interactions for the treatment of Alzheimer's disease. SOJ Pharmacy & Pharmaceutical Sciences. 2014:**1**;1-8. DOI: 10.15226/2374-6866/1/3/00118

[13] Hamada Y, Kiso Y. Recent progress in the drug discovery of non-peptidic BACE1 inhibitors. Expert Opinion on Drug Discovery. 2009:**4**;391-416. DOI: 10.1517/17460440902806377

[14] Hamada Y, Kiso Y. The application of bioisosteres in drug design for novel drug discovery: Focusing on acid protease inhibitors. Expert Opinion on Drug Discovery. 2012:**7**;903-922. DOI: 10.1517/17460441.2012.712513

[15] Hamada Y, Kiso Y. Advances in the identification of β-secretase inhibitors. Expert Opinion on Drug Discovery. 2013:**8**;709-731. DOI: 10.1517/17460441.2013.784267

[16] Nguyen J-T, Hamada Y, Kimura T, Kiso Y. Design of potent aspartic protease inhibitors to treat various diseases. Archiv der Pharmazie—Chemistry in Life Sciences. 2008:**341**;523-535. DOI: 10.1002/ardp.200700267

[17] Sinha S, Anderson JP, Barbour R, et al. Purification and cloning of amyloid precursor protein β-secretase from human brain. Nature. 1999:**402**;537-540. DOI: 10.1038/990114

[18] Ghosh AK, Shin D, Downs D, et al. Design of potent inhibitors for human brain memapsin 2 (β-secretase). Journal of the American Chemical Society. 2000:**122**;3522-3523. DOI: 10.1021/ja000300g

[19] Hong L, Koelsch G, Lin X, et al. Structure of the protease domain of memapsin 2 (β-secretase) complexed with inhibitor. Science. 2000:**290**;150-153. DOI: 10.1126/science.290.5489.150

[20] Ghosh AK, Bilcer G, Harwood, C et al. Structure-based design: Potent inhibitors of human brain memapsin 2 (β-secretase). Journal of Medicinal Chemistry. 2001:**44**;2865-2868. DOI: 10.1021/jm0101803

[21] Ghosh AK, Devasamudram T, Hong L, et al. Structure-based design of cycloamide-urethane-derived novel inhibitors of human brain memapsin 2 (β-secretase). Bioorganic & Medicinal Chemistry Letters. 2005:**15**;15-20. DOI: 10.1016/j.bmcl.2004.10.084

[22] Shuto D, Kasai K, Hidaka K, et al. KMI-008, a novel β-secretase inhibitor containing a hydroxymethylcarbonyl isostere as a transition-state mimic: Design and synthesis of substrate-based octapeptides. Bioorganic & Medicinal Chemistry Letters. 2003:**13**;4273-4276. DOI: 10.1016/j.bmcl.2003.09.053

[23] Kimura T, Shuto D, Kasai K, et al. KMI-358 and KMI-370, highly potent and small-sized BACE1 inhibitors containing phenylnorstatine. Bioorganic & Medicinal Chemistry Letters. 2004:**14**;1527-1531. DOI: 10.1016/j.bmcl.2003.12.088

[24] Kimura T, Shuto D, Hamada Y, et al. Design and synthesis of highly active Alzheimer's β-secretase (BACE1) inhibitors, KMI-420 and KMI-429, with enhanced chemical stability. Bioorganic & Medicinal Chemistry Letters. 2005:**15**;211-215. DOI: 10.1016/j.bmcl.2004.09.090

[25] Asai M, Hattori C, Iwata N, et al. The novel β-secretase inhibitor KMI-429 reduces amyloid beta peptide production in amyloid precursor protein transgenic and wild-type mice. Journal of Neurochemistry. 2006:**96**;533-540. DOI: 10.1111/j.1471-4159.2005.03576.x

[26] Kimura T, Hamada Y, Stochaj M, et al. Design and synthesis of potent β-secretase (BACE1) inhibitors with P1′ carboxylic acid bioisostere. Bioorganic & Medicinal Chemistry Letters. 2006:**16**;2380-2386. DOI: 10.1016/j.bmcl.2006.01.108

[27] Hamada Y, Igawa N, Ikari H, et al. β-Secretase inhibitors: Modification at the P_4 position and improvement of inhibitory activity in cultured cells. Bioorganic & Medicinal Chemistry Letters. 2006:**16**;4354-4359. DOI: 10.1016/j.bmcl.2006.05.046

[28] Hamada Y, Abdel-Rahman H, Yamani A, et al. BACE1 inhibitors: Optimization by replacing the P_1' residue with non-acidic moiety. Bioorganic & Medicinal Chemistry Letters. 2008:**18**;1649-1653. DOI: 10.1016/j.bmcl.2008.01.058

[29] Tagad HD, Hamada Y, Nguyen J-T, et al. Design of pentapeptidic BACE1 inhibitors with carboxylic acid bioisosteres at P_1' and P_4 positions. Bioorganic & Medicinal Chemistry. 2010:**18**;3175-3186. DOI: 10.1016/j.bmc.2010.03.032

[30] Tagad HD, Hamada Y, Nguyen J-T, et al. Structure-guided design and synthesis of P_1' position 1-phenylcycloalkylamine-derived pentapeptidic BACE1 inhibitors. Bioorganic & Medicinal Chemistry. 2011:**19**;5238-5246. DOI: 10.1016/j.bmc.2011.07.002

[31] Hamada Y, Ohta H, Miyamoto N, et al. Significance of interaction of BACE1-Arg235 with its ligands and design of BACE1 inhibitors with P_2 pyridine scaffold. Bioorganic & Medicinal Chemistry Letters. 2009:**19**;2435-2439. DOI: 10.1016/j.bmcl.2009.03.049

[32] Hamada Y, Ohta H, Miyamoto N, et al. Novel non-peptidic and small-sized BACE1 inhibitors. Bioorganic & Medicinal Chemistry Letters. 2008:**18**;1654-1658. DOI: 10.1016/j.bmcl.2008.01.056

[33] Hamada Y, Nakanishi T, Suzuki K, et al. Novel BACE1 inhibitors possessing a 5-nitroisophthalic scaffold at the P_2 position. Bioorganic & Medicinal Chemistry Letters. 2012:**22**;4640-4644. DOI: 10.1016/j.bmcl.2012.05.089

[34] Suzuki K, Hamada Y, Nguyen J-T, et al. Novel BACE1 inhibitors with a non-acidic heterocycle at the P_1' position. Bioorganic & Medicinal Chemistry. 2013:**21**;6665-6673. DOI: 10.1016/j.bmc.2013.08.016

[35] Hamada Y, Suzuki K, Nakanishi T, et al. Structure-activity relationship study of BACE1 inhibitors possessing a chelidonic or 2,6-pyridinedicarboxylic scaffold at the P2 position. Bioorganic & Medicinal Chemistry Letters. 2014:**24**;618-623. DOI: 10.1016/j.bmcl.2013.12.007

[36] Imai YN, Inoue Y, Nakanishi I, Kitaura K. Cl–π interactions in protein–ligand complexes. Protein Science. 2008:**17**;1129-1137. DOI: 10.1110/ps.033910.107

Chapter 4

QSRP Prediction of Retention Times of Chlorogenic Acids in Coffee by Bioplastic Evolution

Francisco Torrens and Gloria Castellano

Additional information is available at the end of the chapter

http://dx.doi.org/10.5772/intechopen.68661

Abstract

Caffeoyl-, feruloyl- and dicaffeoylquinic (chlorogenic) acids in infusions from green and medium roasted coffee beans were identified and quantified by reverse phase liquid chromatography. The chromatographic retention times of chlorogenic acids in coffee are modeled by structure-property relationships. Bioplastic evolution is a view in evolution that conjugates the result of acquired features, and relationships that come out between the principles of evolutionary indeterminacy, morphological determination, and natural selection. Here, it is used to invent the coordination index, which is utilized to typify chlorogenic acids chromatographic retention times. The factors utilized to compute the co-ordination index are the standard molar formation enthalpy, molecular bare, and hydrophobic solvent-accessible surface areas, as well as fractal dimensions. The morphological and coordination indices provide strong correlations. Effect of different types of features is analyzed: thermodynamic, geometric, fractal, etc. Properties are molar formation enthalpy, bare molecular surface area, etc., in linear correlation models. Formation enthalpy, etc. distinguish chlorogenic acids molecular structures.

Keywords: biological plastic evolution, morphological index, co-ordination index, formation enthalpy, molecular surface, hydrophobic accessible surface, fractal dimension, solvation parameter model, chlorogenic acid, hydroxycinnamate, coffee, *Coffea*

1. Introduction

Coffee terpenoids, cafestol, kaweol and 16-*O*-methylcafestol, which occur as fatty acid (FA) esters (FAEs), are responsible for the reversible rise in plasma low-density lipoprotein (LDL) cholesterol (CHOL) observed in some populations (*e.g.*, Scandinavia, Italy) [1–3]. High consumption of boiled, unfiltered coffee was linked to risen levels of homocysteine [4, 5], which, along with risen CHOL, is a known risk factor for cardiovascular disease (CVD). Freshly brewed and instant coffee can induce mutagenic cytotoxic effects *in vitro* [6, 7], because of the gradual formation of hydrogen peroxide H_2O_2 in the beverage [8], which is a reactive oxygen species (ROS) capable of damaging biomolecules and membranes [9]. However, a 5.5% rise in plasma antioxidant activity (AOA) was observed in human volunteers after a single intake of brewed coffee, suggesting its *in vivo* AOA [10].

Coffee contains chlorogenic acids (CGAs) with the amounts varying between green (GCBs) and roasted (RCBs) coffee beans [11, 12]. *Via* an LDL oxidation assay, phenolic compounds in coffee showed AOA, which levels varied depending on coffee beans source and roasting degree [13]. 5-*O*-Caffeoylquinic acid (5-CQA) presented anticarcinogenic properties [14, 15] and a protective role *vs.* LDL oxidation in an *ex vivo* animal model [16]. It and other hydroxycinnamates protected *vs.* the oxidation of human LDL particles *in vitro* [17–20], a process playing a role in atherosclerotic-plaques formation and CVD onset. Crozier and co-workers reported on-line high-performance liquid chromatography (HPLC) analysis of phenolic compounds AOA in brewed, paper-filtered coffee [21]. Antioxidant activity and principles of Vietnam bitter tea *Ilex kudingcha* were informed [22]. Purification and HPLC analysis of CQAs from Kudingcha made from *I. kudingcha* were published [23]. Simultaneous qualitation and quantitation of CGAs in Kuding tea were reported *via* ultra-HPLC-diode array detection coupled with linear ion trap-Orbitrap mass spectrometer (UHPLC-DAD-LTQ-Orbitrap-MS) [24]. Preparation, phytochemical investigation and safety evaluation of CGA products were informed from *Eupatorium adenophorum* [25].

The model used in this work is an extension of solvent-dependent conformational analysis program (SCAP) octanol-water model to organic solvents [26]. In earlier publications, SCAP was applied for partition coefficients of porphyrins, phthalocyanines, benzobisthiazoles, fullerenes, acetanilides, local anesthetics [27], lysozyme [28], barbiturates, hydrocarbons [29], polystyrene [30], Fe–S proteins [31], C-nanotubes [32] and D-glucopyranoses [33]. Bioplastic evolution was applied to phenylalcohols, 4-alkylanilines [34], valence-isoelectronic series of aromatics [35], phenylurea herbicides [36, 37], pesticides [38, 39], methylxanthines and cotinine [40, 41]. Quantitative structure-activity/property relationships (QSARs/QSPRs) were applied to isoflavonoids [42] and sesquiterpene lactones [43]. Mucoadhesive polymer hyaluronan, as biodegradable cationic and zwitterionic-drug delivery vehicle, favors transdermal penetration absorption of caffeine (Caff) [44, 45]. The present report describes QSPR analysis and estimation of CGAs chromatographic retention times. The goal of the study is to identify the properties that differentiate CGAs consistent with chromatographic retention times. The work uses the chemical index in CGAs. The aim of this research is the corroboration of the value of the index by its ability to distinguish CGAs, as well as its concern as a prognostic descriptor for retention time evaluated with regard to molar formation enthalpy, molecular bare, and hydrophobic solvent-accessible surface areas, and fractal dimensions. Section2 describes the computational method. Sections3 and 4 illustrate and discuss the calculation results. Finally, Section5 summarizes our conclusions.

2. Computational method

Biological plastic (bioplastic) evolution is a perspective of the process of the evolution of species. It conjugates the result of (1) the acquired characters and (2) relationships between the principles of evolutionary indeterminacy, morphological determination and natural selection in evolutionary biology. The relationship between morphology and functionality in organisms is that morphology is the substance prop of functionality, which is the dynamic result of the former in the circumstance of the interaction between physical environment and living matter. Morphology, functionality, energy outlay and vital viability are equally affected: When a morphology is functional, it accomplishes its work with minimum energy outlay, and the vital viability of the organ or organism is maximum. Counting these ideas includes defining the *functional coordination index* I_c, which is expressed as the relation between the work achieved by morphology T and the characteristic *morphological index* I_m, consistent with:

$$I_c = \frac{T}{I_m} \tag{1}$$

The larger the work T attained by a specific morphology I_m, the larger the I_c. For an organism, I_m was suggested as the relation between the morphological surface area S and body mass W [46]:

$$I_m = \frac{S}{W} \tag{2}$$

The substitution of Eq. (2) in (1) turns out to be:

$$I_c = \frac{W \cdot T}{S} \tag{3}$$

where T is given by its correspondence in classical mechanics:

$$T = W \cdot x \frac{d^2x}{dt^2} \tag{4}$$

Substituting Eq. (4) in (3), it turns out to be:

$$I_c = \frac{W^2 \cdot x}{S} \frac{d^2x}{dt^2} \tag{5}$$

The I_c rises with the following settings: (1) the larger the body mass at equivalent journeyed time or distance, the larger the I_c; (2) the I_c is proportional to the distance journeyed in the smallest achievable period; (3) the lesser body surface area, the larger I_c and the co-ordination between function and morphology needs lesser energy outlay.

Code SCAP is founded on a representation by Hopfinger, parametrized for 1-octanol-water solvents [47]. The conjecture is that one is able to center a *solvation sphere* on all functional groups in the molecule [48]. The overlapping volume $V°$ between the solvation sphere and van der Waals (VdW) spheres of the resting atoms is computed. Code SCAP handles factors for a solvent: (1) n: maximal number of solvent molecules satisfying the solvation sphere; (2) $\Delta g°$: change of Gibbs free energy linked with the removal of one molecule of solvent out of the solvation sphere [49, 50]; (3) R_v: radius of the solvation sphere; (4) V_f: free volume accessible for a solvent molecule in the solvation sphere [51]. In the solvation sphere, a fraction of its volume keeps out solvent molecules. The volume is made of VdW volume of the functional group at which the sphere is centered and a volume standing for functional groups connected to the central one. This volume is symbolized by a set of cylinders. The difference between the total volume of the solvation sphere and that prohibited for the solvent molecules stands for the volume V' that is accessible for n solvent molecules. The V_f is computed by $V_f = V'/n - V_s$. The variation of free energy linked to the removal of every solvent molecule out of the solvation sphere of a functional group R turns out to be: $\Delta G_R{}^\circ = n\Delta g^\circ(1 - V^\circ/V')$, and the free energy of solvation of molecule results: $\Delta G_{solv}{}^\circ = -\Sigma_{R=1}{}^N \Delta G_R{}^\circ$. The 1-octanol-water partition coefficient P results:

$$RT\ \ln P = \Delta G^o_{solv}(\text{water}) - \Delta G^o_{solv}(1 - \text{octanol}) \tag{6}$$

at a certain temperature T got as 298 K, where R is the gas constant, and $\Delta G_{solv}{}^\circ$(1-octanol) and $\Delta G_{solv}{}^\circ$(water), in kJ·mol^{-1}, are the standard-state Gibbs solvation free energies. Extending SCAP for other solvents, the 1-octanol factors were customized considering the result of solvent permittivity and molecular volume. For a general solvent, the maximal number of solvent molecules permitted packing the solvation sphere is connected to the molecular volume of the solvent by:

$$n_s = n_o \left(\frac{V_s}{V_o}\right)^{\frac{\log \frac{n_o}{n_w}}{\log \frac{V_o}{V_w}}} \tag{7}$$

where V_o, V_w and V_s are the molecular volumes of 1-octanol, water and the general solvent. The n_o, n_w and n_s are the maximal numbers of molecules of 1-octanol, water, and the general solvent allowed packing the solvation sphere. The change in the standard Gibbs free energy connected to the removal of one molecule of solvent out of the solvation sphere, $\Delta g_s{}^\circ$, is computed *via* the extended Born equation:

$$\Delta g^o_s = \Delta g^o_o \frac{1 - \frac{1}{\varepsilon_s}}{1 - \frac{1}{\varepsilon_o}} = \Delta g^o_o \frac{\varepsilon_o(\varepsilon_s - 1)}{\varepsilon_s(\varepsilon_o - 1)} \tag{8}$$

where $\Delta g_s{}^\circ$ is Δg° for 1-octanol, and ε_o and ε_s are the permittivities of 1-octanol and the general solvent. The radius of the solvation sphere results connected to the molecular volume of the solvent molecule by:

$$R_{v,s} = R_{v,o}\left(\frac{V_s}{V_o}\right)^{\frac{1}{3}} \tag{9}$$

where $R_{v,o}$ is R_v in 1-octanol. The free volume accessible for a solvent molecule in the solvation sphere is:

$$V_{f,s} = V_{f,o}\frac{V_s}{V_o} \tag{10}$$

where $V_{f,o}$ is V_f in 1-octanol.

The models were obtained *via* multiple linear regression (MLR). Correlation coefficient r was used as the calibration function of the regression models, together with standard deviation s, variance ratio F, prediction error sum of squares (PESS), mean absolute percentage error (MAPE) and approximation error variance (AEV). Statistics r, s and F were calculated with Microsoft Excel 2016, and PESS, MAPE and AEV, with Knowledge Miner for Excel. Our codes SCAP and TOPO [52] are available from the authors at the Internet (torrens@uv.es) and are free for academics.

3. Calculation results

For nine CGAs, *viz.* 3/4/5-*O*-caffeoyl- (CQAs), 3/4/5-*O*-feruloyl- (FQAs) and 3, 4/3, 5/4, 5-*O*-dicaffeoylquinic (diCQAs) acids (*cf.* **Figure 1**), reverse phase (RP) HPLC retention times R_t were taken from Crozier and co-workers.

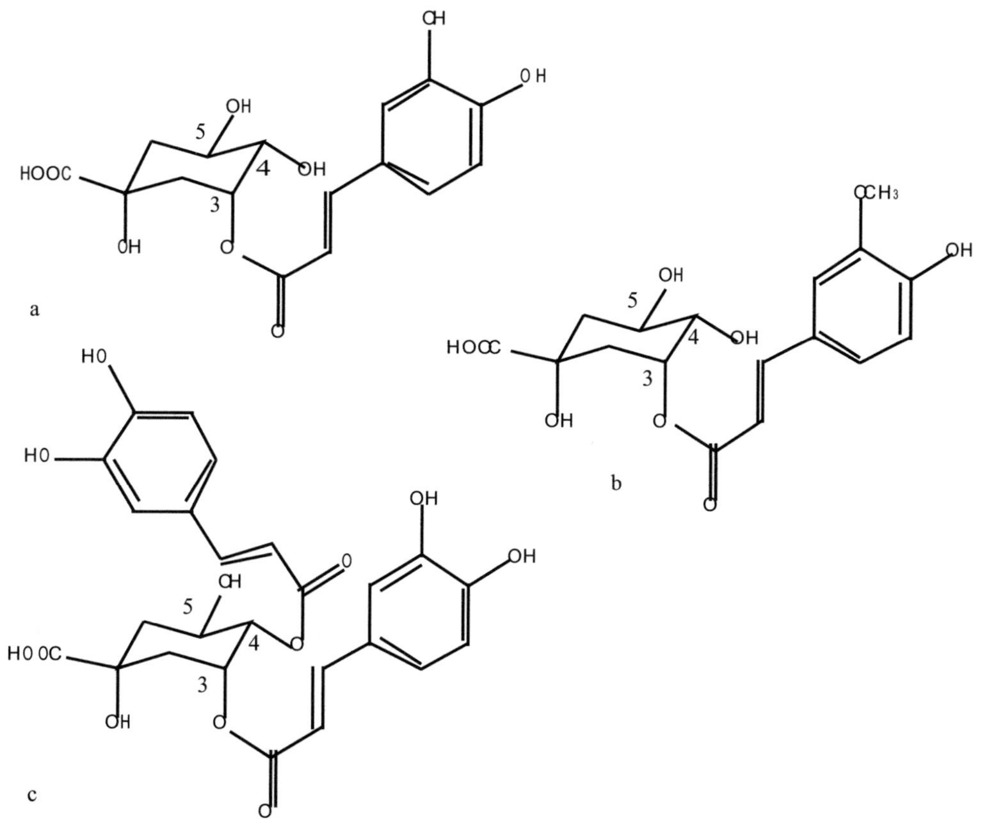

Figure 1. Structures: (a) 3-*O*-caffeoylquinic; (b) 3-*O*-feruloylquinic; (c) 3,4-*O*-dicaffeoylquinic acids.

The 3-CQA was taken as *reference* retention time R_t° because of least R_t (*cf.* **Table 1**). Ratios (R_t-R_t°)/R_t° were calculated. Standard molar formation enthalpy was computed with code MOPAC-AM1 [53]. Molecular bare S and hydrophobic solvent (water)-accessible (HBAS) surface areas, fractal dimension D and this averaged for non-buried atoms D' were calculated with our code TOPO.

The use of the co-ordination index in the chemical description of molecules needs to change variables T, S and W [Eq. (3)]: T is redescribed as minus the standard molar formation enthalpy (kJ·mol^{-1}), S is the molecular surface area (Å^2), and W is the molecular mass (g·mol^{-1}). Chemical indices for CGAs characterization (*cf.* **Table 2**) show that I_m remains constant, while I_c rises with W.

Indices variation for CGAs *vs.* molecular weight W (*cf.* **Figure 2**) shows that most points collapse for CQAs, FQAs and diCQAs, every group with three isomers. The only descriptor that remains almost constant is I_m. Descriptors more sensitive to W decay: $I_c > T > S > I_m$.

Variations of (R_t-R_t°)/R_t° *vs.* morphological index I_m show fit; the regression turns out to be:

$$\frac{R_t - R_t^o}{R_t^o} = 43.7 - 45.3 I_m \quad (11)$$

$n = 9, r = 0.808, s = 1.047, F = 13.2.$

PESS = 0.3972, MAPE = 36.39%, AEV = 0.3472.

Molecule	R_t (min)	R_t-R_t° (min)	(R_t-R_t°)/R_t°	ΔH_f° (kJ·mol^{-1})[a]	HBAS (Å^2)[b]	D[c]	D'[d]
1. 3-*O*-Caffeoylquinic acid (3-CQA)	9.0	0.0	0.000	−1545.4	218.42	1.390	1.480
2. 4-*O*-Caffeoylquinic acid (4-CQA)	13.6	4.6	0.511	−1550.0	241.44	1.375	1.498
3. 5-*O*-Caffeoylquinic acid (5-CQA)	15.4	6.4	0.711	−1570.5	238.38	1.387	1.458
4. 3-*O*-Feruloylquinic acid (3-FQA)	16.2	7.2	0.800	−1519.5	281.61	1.410	1.490
5. 4-*O*-Feruloylquinic acid (4-FQA)	23.6	14.6	1.622	−1524.2	305.78	1.383	1.511
6. 5-*O*-Feruloylquinic acid (5-FQA)	27.3	18.3	2.033	−1541.4	301.12	1.395	1.476
7. 3,4-*O*-Dicaffeoylquinic acid (3,4-diCQA)	41.9	32.9	3.656	−1839.7	296.65	1.445	1.534
8. 3,5-*O*-Dicaffeoylquinic acid (3,5-diCQA)	44.7	35.7	3.967	−1865.2	322.59	1.453	1.562
9. 4,5-*O*-Dicaffeoylquinic acid (4,5-diCQA)	49.1	40.1	4.456	−1867.2	308.05	1.434	1.500

[a]Standard molar formation enthalpy calculated with MOPAC-AM1.
[b]HBAS: hydrophobic water-accessible surface area.
[c]D: molecular fractal dimension.
[d]D': molecular fractal dimension averaged for non-buried atoms.

Table 1. Retention time, formation enthalpy and fractal dimensions for chlorogenic acids.

Molecule	W (g·mol^{-1})[a]	T (kJ·mol^{-1})[b]	S (Å^2)[c]	I_m (mol·Å^2·g^{-1})[d]	I_c (kJ·g·mol^{-2}·Å^{-2})[e]
3-*O*-Caffeoylquinic acid	354	1545.4	328.77	0.929	1664.0
4-*O*-Caffeoylquinic acid	354	1550.0	329.08	0.930	1667.4
5-*O*-Caffeoylquinic acid	354	1570.5	335.65	0.948	1656.4
3-*O*-Feruloylquinic acid	368	1519.5	345.85	0.940	1616.8
4-*O*-Feruloylquinic acid	368	1524.2	346.24	0.941	1620.0
5-*O*-Feruloylquinic acid	368	1541.4	351.14	0.954	1615.4
3,4-*O*-Dicaffeoylquinic acid	516	1839.7	459.17	0.890	2067.4
3,5-*O*-Dicaffeoylquinic acid	516	1865.2	464.60	0.900	2071.6
4,5-*O*-Dicaffeoylquinic acid	516	1867.2	447.14	0.867	2154.8

[a]W: molecular weight (g·mol^{-1}).
[b]T: minus standard formation enthalpy (kJ·mol^{-1}).
[c]S: molecular surface area (Å^2).
[d]I_m: morphological index (mol·Å^2·g^{-1}).
[e]I_c: co-ordination index (kJ·g·mol^{-2}·Å^{-2}).

Table 2. Bioplastic evolution indices for chlorogenic acids.

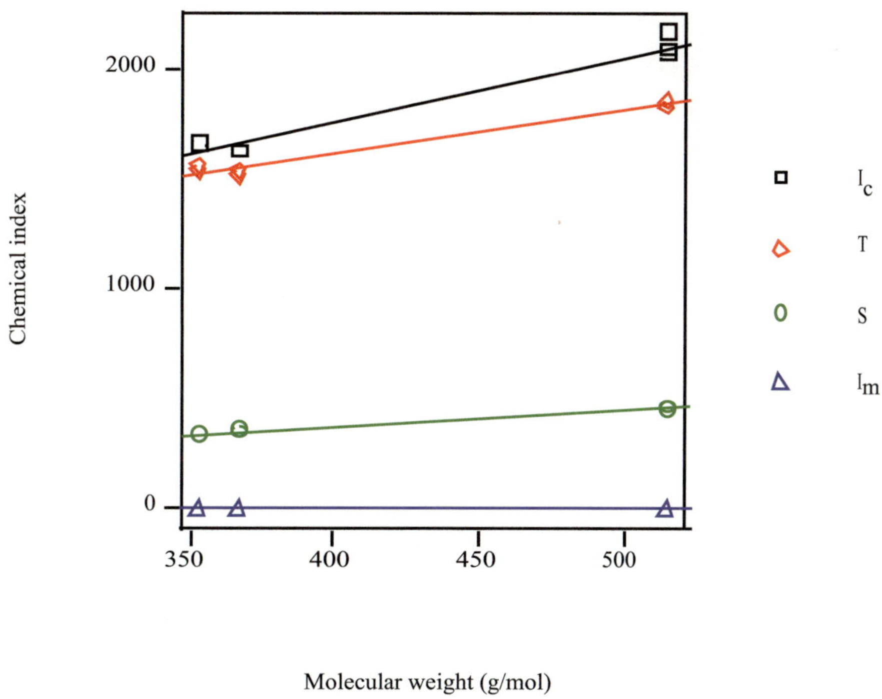

Figure 2. Variation of chemical indices for CGAs *vs*. molecular weight: $y = 589+2.92x$; $y = 817+2.01x$; $y = 64.7+0.761x$; $y=1.07–0.000346x$.

where MAPE is 36.39% and AEV, 0.3472. The use of coordination index I_c improves the model:

$$\frac{R_t - R_t^o}{R_t^o} = -9.70 + 0.00651 I_c \tag{12}$$

$n = 9$, $r = 0.906$, $s = 0.754$, $F = 31.9$.

PESS = 0.2038, MAPE = 25.80%, AEV = 0.1800.

and AEV decays by 48%. The utilization of the standard molar formation enthalpy betters the fit:

$$\frac{R_t - R_t^o}{R_t^o} = -13.8 - 0.00959 \Delta H_f^o \tag{13}$$

$n = 9$, $r = 0.916$, $s = 0.714$, $F = 36.3$.

PESS = 0.1857, MAPE = 26.15%, AEV = 0.1616.

and AEV drops by 53%. The application of the bare molecular surface area S improves the model:

$$\frac{R_t - R_t^o}{R_t^o} = -8.09 + 0.0266 S \tag{14}$$

$n = 9$, $r = 0.950$, $s = 0.556$, $F = 64.5$.

PESS = 0.1232, MAPE = 22.11%, AEV = 0.0979.

and AEV decreases by 72%.

The inclusion of the hydrophobic solvent (water)-accessible surface area HBAS improves the fit:

$$\frac{R_t - R_t^o}{R_t^o} = -12.3 + 0.00472 I_c + 0.0210 \text{HBAS} \tag{15}$$

$n = 9$, $r = 0.988$, $s = 0.291$, $F = 127.5$.

PESS = 0.0365, MAPE = 9.86%, AEV = 0.0230.

and AEV decays by 93%. The fractal dimension averaged for *non-buried* atoms D' betters the fit:

$$\frac{R_t - R_t^o}{R_t^o} = -2.57 - 0.00801 \Delta H_f^o + 0.0229 \text{HBAS} - 10.0 D' \tag{16}$$

$n = 9$, $r = 0.995$, $s = 0.204$, $F = 174.7$.

PESS = 0.0141, MAPE = 6.63%, AEV = 0.0095.

and AEV decreases by 97%. The incorporation of the fractal dimension D improves the model:

$$\frac{R_t - R_t^o}{R_t^o} = 7.86 - 0.00899\Delta H_f^o + 0.0257\text{HBAS} - 7.80D - 11.3D' \tag{17}$$

$n = 9, r = 0.998, s = 0.154, F = 232.2.$

$\text{PESS} = 0.0091, \text{MAPE} = 5.16\%, \text{AEV} = 0.0058.$

and AEV drops by 98%. The inclusion of the bare molecular surface area S betters the correlation:

$$\frac{R_t - R_t^o}{R_t^o} = 11.0 - 0.00776\Delta H_f^o + 0.00606S + 0.0222\text{HBAS} - 11.2D - 9.71D' \tag{18}$$

$n = 9, r = 0.999, s = 0.136, F = 239.1.$

$\text{PESS} = 0.0060, \text{MAPE} = 4.00\%, \text{AEV} = 0.0038.$

and AEV decays by 99%. The best non-linear models do not improve the correlation. Additional fitting parameters were tested: molecular dipole moment, weight, volume, globularity, rugosity, hydrophilic and total solvent-accessible surfaces, accessibility and fractal dimension for external atoms minus fractal index (D'–D). Notwithstanding, the results do not improve Eqs. (12)–(18).

4. Discussion

Food effects on health rightly worry consumers. Mass media tend to satisfy the permanent question, and physicians must face many queries from the persons that come to consult them. Information sources are scattered in many scientific journals, and a few domains exist that be so dispersed in different databases international journals. Information circulates badly, critical syntheses are rare, and an important passivity exists in knowledge transmission. Because of the great interest devoted to their health, consumers are receptive to all new accounts that concern food. Mass media know it and reply in a simplified way *via* all new data, where the impact will be proportional to novelty character. Results of scientific studies must be interpreted *vs.* experimental conditions, transmitting them without nuance finish in a misinformation and myths creation. Cafés popularization during the nineteenth century, replacing beer bars, decreased alcohol consumption during working days, improving health and safety at work. Daily coffee is something to which many people cannot renounce. It does not matter if it is consumed to increase energy or enjoy it in the company but one thing is clear: The taste and quality are important factors. In order to guarantee the best taste and quality, many RCB companies place their trust in good-quality equipment. Many physiological properties either favorable or unfavorable to health were attributed to coffee [54]. Some are exact, some other, mistaken. Errors come from two causes: tradition and experimental results interpretation. As coffee physiological effects frequently entail by subjective observations, and their intensity is variable from one person to another, early generalizations were made, which led to definitive takings of the position that entertain public misinformation. According to some studies, to drink three or four cups of coffee per day presents positive effects for health. It is indifferent that it be decaffeinated or not. Besides Caff and flavorings, coffee is rich in antioxidants, responsible for this be so healthy. Decaffeinated and GCBs would be more healthy than RCBs coffee. Caffeine is more concentrated in tea leaves than in GCB/RCBs. However, more caffeine exists in a coffee than in a tea drink because of the different preparation methods. Molecule Caff acts by impeding adenosine $A_{1/2A}$ receptors ($A_{1/2A}R$), pointing to that some A_1Rs are tonically more active. Mice were made with a targeted disturbance of the second coding exon of A_1R ($A_1R^{-/-}$) [55]. They raised and increased mass as normal, and presented a usual heart speed, blood pressure, and body temperature. In the majority of behavioral experiments, they resulted alike $A_1R^{+/+}$ but $A_1R^{-/-}$ mice presented signals of risen nervousness. Electrophysiological footages from pieces of the hippocampus showed that the inhibition arbitrated by adenosine and increase arbitrated by theophylline of excitatory glutamatergic neurotransmission resulted put to an end in $A_1R^{-/-}$ mice. In $A_1R^{+/-}$ mice, adenosine activity was halved, as resulted in the figure of A_1Rs. In $A_1R^{-/-}$ mice, the painkilling consequence of intrathecal adenosine resulted misplaced, and thermal hyperalgesia was shown, but morphine painkilling result was whole. The decay of neuronal potency on hypoxia decreased in pieces of the hippocampus and brainstem, and working revival after hypoxia decreased. The A_1Rs do not perform a fundamental position throughout development and, though they affect synaptic potency, they perform an additional position in usual physiology. However, beneath pathophysiological circumstances (*e.g.*, noxious incentive, O_2 lack), these receptors result significant. Coffee abuse turns people weaker. *Taking high Caff doses* per day for a long time *turns people more sensitive to pain* and hypoxias [56]. Caffeine is stimulant and counterproductive. Not all Caff effects are negative; for example, it is fine for vasodilating. When a premature newborn baby suffers from apnoeas (breath suspensions), administering him Caff improves lung functionality. Caffeine is also administered to patients suffering from asthma because it helps them to bronchodilate. However, all at moderate doses, as high Caff doses present the opposite effect. Another contradiction is that although coffee helps to digest, Caff is a gastric irritant because it increases the production of saliva, HCl, and substances that are released with gastric juices. It is counter-indicated in the case of ulcers and gastritis, but it presents a positive effect in the cases of gallstones as it reduces 40% the risk of suffering from them. In most cases, Caff is more negative than positive for health, although at small doses, which is what people normally take (*e.g.*, one cup of coffee in the morning, another after lunch), it presents beneficial effects as it acts as a stimulant. In pregnant women, Caff effect is higher and it is harder for them to eliminate it. Those born of women that during pregnancy drink a lot of coffee, in the first hours of life, present a small abstinence-syndrome symptomatology. It is because Caff causes dependence. When one does not drink coffee (if he habitually does it), he suffers from a headache, fatigue sensation, apathy, irritability, marked sleepiness, etc. Symptoms disappear when one drinks it again. Although one cannot label it as an addictive substance, it is considered doping, and high-competition athletes cannot drink it because it is considered a psychoactive substance that stimulates resistance and muscular strength. Caffeine is also present in tea, colas (*e.g.*, Coca-Cola®) and cocoa, although in the last, the quantity is derisory. The quantity of Caff that one consumes also depends on how the coffee is served, type of coffee or tea, etc.; for example, green (GT) presents 40% less Caff than black tea (BT) because the latter is oxidized (*fermented*), which makes that Caff come out easier. Maximum recommended quantity of Caff is 500mg/day. Caffeine is a drug model because it is one of the most studied medicines. It is the world's most widely consumed psychoactive drug. Theobromine may serve as a lead compound for novel drugs development. Analysis of Caff, its metabolites and phenolic compounds (CGAs) in foods, beverages, human plasma and urine is difficult because of the complex food, blood, and urine matrixes. Despite their progressive destruction during roasting, substantial amounts of CGAs survive to be extracted into domestic brews and instant coffee, and for many consumers, the beverage must be major dietary source of not only CGAs but also other antioxidants.

One of the important applications of QSAR/QSPR models is to fill data gaps, by predicting a given response property or activity from known molecular features, or physicochemical and physiological properties of new compounds, which might not be experimentally tested. The performance of a model should be evaluated based on predictions quality from the test and not from the training set, in order to obviate any overfitting problem. The use of phenomenological methods, for example, QSAR/QSPR, is restricted by the insufficient accuracy of final digits. A quantum-mechanical consideration of additive models showed that in most phenomenological approaches, systematic error is composed of two methodical errors: the same contribution of formally identical fragments and the inclusion of small molecules in training set. Two ways to improve models prognostic capabilities are: (1) compensation by introducing additional terms and (2) elimination of models systematic error. Building a model, Occam's razor (principle of maximal parsimony) philosophical approach should be used, that is, fit the least complex (most parsimonious) model that could correctly describe training data. The simpler the model, the better the generalization one is going to find.

A study was made of the relations between retention times obtained by RP-HPLC chromatography for a group of CGAs. *Via* multivariate linear regression, the corresponding molecular functions were obtained, which were selected based on their respective statistical parameters. Regression analysis of molecular functions showed a forecast of experimental elution sequence for CGAs. In order to predict experimental elution sequence in CGAs, 1–5-variable models were necessary in which the appearance of coordination index, morphological indicator, molar formation enthalpy, bare surface area S, hydrophobic water-accessible surface HBAS, D and D' reveals thermodynamic, geometric and fractal analyses importance in the studied property, allowing the use of such equations in forecasting property value. Molecular structures may be differentiated even in other phenolic compounds not included in the series, in brewed, paper-filtered coffee.

The QSPR linear models explaining the variation of chromatographic relative retention time *vs.* morphological I_{m} and coordination I_{c} indices show a negative correlation with I_{m} [Eq. (11)] and a positive association with I_{c} [Eq. (12)]. The best model is for index I_{c} [Eq. (12)], according to all statistics: correlation coefficient, standard deviation, variance ratio, prediction error sum of squares, mean absolute percentage error and approximation error variance.

Thermodynamic indices were tried in order to improve the model. The molar formation enthalpy negatively correlates with the relative retention time and betters the fit [Eq. (13)].

Geometric descriptors were assayed in order to improve the fit. The molecular surface positively correlates with the relative retention time and betters the model [Eq. (14)]. The inclusion of the hydrophobic accessible surface presents a positive correlation with the relative retention time and improves the fit [Eq. (15)]. Notice that in this equation, index I_c shows a positive correlation, in agreement with Eq. (12).

Topological indices were tried in order to improve the model. The incorporation of the fractal dimension averaged for external (*non-buried*) atoms negatively correlates with the relative retention time and betters the fit [Eq. (16)]. In this equation, index ΔH_f° shows a negative correlation, in agreement with Eq. (13), and index HBAS presents positive association, in agreement with Eq. (15). The inclusion of the fractal dimension negatively correlates with the relative retention time and improves the fit [Eq. (17)]. In this equation, index ΔH_f° presents negative correlation, in agreement with Eqs. (13) and (16), index HBAS presents positive association, in agreement with Eqs. (15) and (16), and index D' presents negative correlation, in agreement with Eq. (16). The inclusion of index S positively correlates with the relative retention time and improves the fit [Eq. (18)], in agreement with Eq. (14). In Eq. (18), index ΔH_f° presents negative correlation, in agreement with Eqs. (13), (16) and (17), index HBAS shows a positive association, in agreement with Eqs. (15)–(17), index D presents negative correlation, in agreement with Eq. (17), and index D' shows a negative association, in agreement with Eqs. (16) and (17).

5. Conclusion

From the present results and discussion, the following conclusions can be drawn.

1. The objective of this study was to develop structure-property relationships for the qualitative and quantitative prediction of the reverse phase liquid chromatographic retention times of CGAs. It is hoped that the results of the present work increase scientific knowledge in the field of the relation prediction of chlorogenic acids in coffee. Program SCAP permits the Gibbs free energies of solvation (hydration) and partition coefficients that illustrate that for a certain atom, the solvation energies and partition coefficients result responsive to the occurrence in the molecule of some other atoms and groups.

2. The factors necessary to compute the co-ordination index result in the standard molar formation enthalpy, molecular mass and surface area.

3. Linear correlation models resulted for chromatographic retention times. The morphological and coordination indices provided strong multivariable linear regression equations for chromatographic retention. The trend between the coordination index and molecular weight points not only to a homogeneous molecular structure of chlorogenic acids but also to the ability to predict and tailor their properties. The latter is non-trivial in the analysis of chlorogenic acids and phenolic compounds in foods, beverages, human plasma, and urine because of the complex food, blood and urine matrixes.

4. The effect of different types of features was analyzed: thermodynamic, geometric, fractal, etc. The molar formation enthalpy, bare molecular and hydrophobic solvent-accessible surface areas, fractal dimensions, etc. distinguished chlorogenic acids in linear fits.

http://dx.doi.org/10.5772/intechopen.68661

Acknowledgements

The authors acknowledge support from Generalitat Valenciana (Project No. PROMETEO/2016/094) and Universidad Católica de Valencia *San Vicente Mártir* (Projects Nos. PRUCV/2015/617 and 2017).

Author details

Francisco Torrens[1]* and Gloria Castellano[2]

*Address all correspondence to: torrens@uv.es

1 University Institute for Molecular Science, Universitat de València-ICMol, Valencia, Spain

2 Department of Experimental Sciences and Mathematics, Faculty of Veterinary and Experimental Sciences, Valencia Catholic University Saint Vincent Martyr, Valencia, Spain

References

[1] Urgert R, Meyboom S, Kuilman M. Comparison of effect of cafetiere and filtered coffee on serum concentrations of liver aminotransferases and lipids: Six month randomised controlled trial. British Medical Journal. 1996;**313**:1362–1366

[2] Urgert R, Weusten-van der Woum MP, Hovenier R. Diterpenes from coffee beans decrease serum levels of lipoprotein (a) in humans: Results from four randomised controlled trials. European Journal of Clinical Nutrition. 1997;**51**:431–436

[3] Urgert R, Katan MB. The cholesterol-raising factor from coffee beans. Annual Review of Nutrition. 1997;**17**:305–324

[4] Urgert R, van Vliet T, Zock PL, Katan MB. Heavy coffee consumption and plasma homocysteine: A randomised controlled trial in healthy volunteers. The American Journal of Clinical Nutrition. 2000;**72**:1107–1110

[5] Verhoef P, Pasman WJ, van Vliet T, Urgert R, Katan MB. Contribution of caffeine to the homocysteine-raising effect of coffee: A randomised controlled trial in humans. The American Journal of Clinical Nutrition. 2002;**76**:1244–1248

[6] Nagao M, Fujita Y, Wakabayashi K, Nukaya H, Kosuge T, Sugimura T. Mutagens in coffee and other beverages. Environmental Health Perspectives. 1986;**67**:89–91

[7] Itagaki SK, Kobayashi T, Kitagawa Y, Iwata S, Nukaya H, Tsuji K. Cytotoxicity of coffee in human intestinal cells *in vitro* and its inhibition by peroxidase. Toxicology In Vitro. 1992;**6**:417–421

[8] Stadler RH, Turesky RJ, Müller O, Markovic J, Leong-Morgenthaler PM. The inhibitory effects of coffee on radical-mediated oxidation and mutagenicity. Mutation Research. 1994;**308**:177–190

[9] Shah AM, Channon KM. Free radicals and redox signalling in cardiovascular disease. Heart. 2004;**90**:486–487

[10] Natella F, Nardini M, Giannetti I, Dattilo C, Scaccini C. Coffee drinking influences plasma antioxidant capacity in humans. Journal of Agricultural and Food Chemistry. 2002;**50**:6211–6216

[11] Clifford MN. Chlorogenic acids and other cinnamates: Nature, occurrence and dietary burden. Journal of the Science of Food and Agriculture. 1999;**79**:362–372

[12] Clifford MN. Chlorogenic acids and other cinnamates: Nature, occurrence, dietary burden, absorption and metabolism. Journal of the Science of Food and Agriculture. 2000;**80**: 1033–1043

[13] Richelle M, Tavazzi I, Offord E. Comparison of the antioxidant activity of commonly consumed polyphenolic beverages (coffee, cocoa, and tea) prepared per cup serving. Journal of Agricultural and Food Chemistry. 2001;**49**:3438–3442

[14] Huang MT, Smart RC, Wong CQ, Connay AH. Inhibitory effect of curcumin, chlorogenic acid, caffeic acid and ferulic acid on tumor promotion in mouse skin by 12-*O*-tetradecanoylphorbol-13-acetate. Cancer Research. 1988;**48**:5941–5946

[15] Tanaga T, Kojima T, Kawamori T, Wang A, Suzui M, Okamoto K, Mori H. Inhibition of 4-nitroquinoline-1-oxide-induced rat tongue carcinogenesis by the naturally occurring plant phenolics caffeic, ellagic, chlorogenic and ferulic acids. Carcinogenesis. 1993;**14**:1321–1325

[16] Castelluccio C, Paganga G, Melikian N, Bolwell GP, Pridham J, Sampson J, Rice-Evans C. Antioxidant potential of intermediates in phenylpropanoid metabolism in higher plants. FEBS Letters. 1995;**368**:188–192

[17] Meyer AS, Donovan JL, Pearson DA, Waterhouse AL, Frankel EN. Fruit hydroxycinnamic acids inhibit human low-density lipoprotein oxidation in-vitro. Journal of Agricultural and Food Chemistry. 1998;**46**:1783–1787

[18] Moon JH, Terao J. Antioxidant activity of caffeic acid and dihydrocaffeic acid in lard and human low-density lipoprotein. Journal of Agricultural and Food Chemistry. 1998;**46**:5062–5065

[19] Morton LW, Caccetta RAA, Puddey IB, Croft KD. Chemistry and biological effects of dietary phenolic compounds: Relevance to cardiovascular disease. Clinical and Experimental Pharmacology and Physiology. 2000;**27**:152–159

[20] Andreasen MF, Landbo AK, Christensen LP, Hansen Å, Meyer AS. Antioxidant effects of phenolic rye (*Secale cereale* L.) extracts, monomeric hydroxycinnamates, and ferulic acid dehydrodimers on human low-density lipoproteins. Journal of Agricultural and Food Chemistry. 2001;**49**:4090–4096

[21] Stalmach A, Mullen W, Nagai C, Crozier, A. On-line HPLC analysis of the antioxidant activity of phenolic compounds in brewed, paper-filtered coffee. Brazilian Journal of Plant Physiology. 2006;**18**:253–262

[22] Thuong PT, Su ND, Ngoc TM, Hung TM, Dang NH, Thuan ND, Bae KH, Oh WK. Antioxidant activity and principles of Vietnam bitter tea *Ilex kudingcha*. Food Chemistry. 2009;**113**:139–145

[23] Wang QC, Zhang X, Zhang WQ, Sun XQ, Hu B, Sun Y, Zeng XX. Purification and HPLC analysis of caffeoylquinic acids from Kudingcha made from *Ilex kudingcha* C. J. Tseng. Journal of Food Science. 2013;**34**:119–122

[24] Che Y, Wang Z, Zhu Z, Ma Y, Zhang Y, Gu W, Zhang J, Rao G. Simultaneous qualitation and quantitation of chlorogenic acids in Kuding tea using ultra-high-performance liquid chromatography–diode array detection coupled with linear ion trap–Orbitrap mass spectrometer. Molecules 2016;**21**:1728-1–14

[25] Liu, B, Cao, L, Zhang, L, Yuan, X, Zhao, B. Preparation, phytochemical investigation, and safety evaluation of chlorogenic acid products from *Eupatorium adenophorum*. Molecules 2017;**22**:67-1–12

[26] Torrens F, Sánchez-Marín J, Nebot-Gil I. Universal model for the calculation of all organic solvent–water partition coefficients. Journal of Chromatography A. 1998;**827**:345–358

[27] Torrens F. Universal organic solvent-water partition coefficient model. Journal of Chemical Information and Modeling. 2000;**40**:236–240

[28] Torrens F. Calculation of partition coefficient and hydrophobic moment of the secondary structure of lysozyme. Journal of Chromatography A. 2001;**908**:215–221

[29] Torrens F. Free energy of solvation and partition coefficients in methanol–water binary mixtures. Chromatographia. 2001;**53**:S199-S203

[30] Torrens F, Soria V. Stationary-mobile phase distribution coefficient for polystyrene standards. Separation Science and Technology. 2002;**37**:1653–1665

[31] Torrens F. Calculation of organic solvent–water partition coefficients of iron–sulfur protein models. Polyhedron. 2002;**21**:1357–1361

[32] Torrens F. Calculation of solvents and co-solvents of single-wall carbon nanotubes: Cyclopyranoses. Nanotechnology. 2005;**16**:S181-S189

[33] Torrens F, Castellano G. (Co-)solvent selection for single-wall carbon nanotubes: *Best* solvents, acids, superacids and guest–host inclusion complexes. Nanoscale. 2011;**3**:2494–2510

[34] Torrens F. A new chemical index inspired by biological plastic evolution. Indian Journal of Chemistry Sec A. 2003;**42**:1258–1263

[35] Torrens F. A chemical index inspired by biological plastic evolution: Valence-isoelectronic series of aromatics. Journal of Chemical Information and Modeling. 2004;**44**:575–581

[36] Torrens F, Castellano G. QSPR prediction of retention times of phenylurea herbicides by biological plastic evolution. Current Drug Safety. 2012;**7**:262–268

[37] Torrens F, Castellano G. Molecular clustering of phenylurea herbicides: Comparison with sulphonylureas, pesticides and persistent organic pollutants. Evolving Trends in Engineering and Technology. 2014;**1**:29–52

[38] Torrens F, Castellano G. QSPR prediction of chromatographic retention times of pesticides: Partition and fractal indices. Journal of Environmental Science and Health, Part B. 2014;**49**:400–407

[39] Torrens F, Castellano G. Molecular classification of pesticides including persistent organic pollutants, phenylurea and sulphonylurea herbicides. Molecules. 2014;**19**:7388–7414

[40] Torrens F, Castellano G. QSPR prediction of retention times of methylxanthines and cotinine by bioplastic evolution. International Journal of Quantitative Structure-Property Relationships. in press

[41] Torrens F, Castellano G. Molecular classification of caffeine, its metabolites and nicotine metabolite. In: Ul-Haq Z, Madura JD, editors. Frontiers in Computational Chemistry. Hilversum (Holland): Bentham; Vol. 4, in press

[42] Castellano G, Torrens F. Quantitative structure–antioxidant activity models of isoflavonoids: A theoretical study. International Journal of Molecular Sciences. 2015;**16**:12891–12906

[43] Castellano G, Redondo L, Torrens F. QSAR of natural sesquiterpene lactones as inhibitors of Myb-dependent gene expression. Phytochemistry, submitted for publication

[44] Torrens F, Castellano G. Mucoadhesive polymer hyaluronan as biodegradable cationic/zwitterionic-drug delivery vehicle. ADMET DMPK. 2014;**2**:235–247

[45] Torrens F, Castellano G. Computational study of nanosized drug delivery from cyclodextrins, crown ethers and hyaluronan in pharmaceutical formulations. Current Topics in Medicinal Chemistry. 2015;**15**:1901–1913

[46] Ruíz-Bustos A. La Evolución Plástica. Granada (Spain): Andalucía; 1994

[47] Hopfinger AJ. Polymer-solvent interactions for homopolypeptides in aqueous solution. Macromolecules. 1971;**4**:731–737

[48] Hopfinger AJ, Battershell RD. Application of SCAP to drug design: 1. Prediction of octanol–water partition coefficients using solvent-dependent conformational analyses. Journal of Medicinal Chemistry. 1976;**19**:569–573

[49] Gibson KD, Scheraga HA. Minimization of polypeptide energy. I. Preliminary structures of bovine pancreatic ribonuclease S-peptide. Proceedings of the National Academy of Sciences of the United States of America. 1967;**58**:420–427

[50] Rekker RF. The Hydrophobic Fragmental Constant. Amsterdam (The Netherlands): Elsevier; 1976

[51] Pascal P. Program SCAP. Nancy (France): Université Henry Poincaré-Nancy I; 1991

[52] Torrens F. Characterizing cavity-like spaces in active-site models of zeolites. Computational Materials Science. 2003;**27**:96–101

[53] Dewar MJS, Zoebisch EG, Healy EF, Stewart JJP. AM1: A new general purpose quantum mechanical model. Journal of the American Chemical Society. 1985;**107**:3902–3909

[54] Debry G. Le Café: Sa Composition, sa Consommation, ses Incidences sur la Santé. Centre de Nutrition Humaine Monographie No. 1. Paris (France): Communications Economiques et Sociales; 1990

[55] Johansson B, Halldner L, Dunwiddie TV, Masino SA, Poelchen W, Giménez-Llort L, Escorihuela RM, Fernández-Teruel A, Wiesenfeld-Hallin Z, Xu XJ, Hårdemark A, Betsholtz C, Herlenius E, Fredholm BB. Hyperalgesia, anxiety, and decreased hypoxic neuroprotection in mice lacking the adenosine A_1 receptor. Proceedings of the National Academy of Sciences of the United States of America. 2001;**98**:9407–9412

[56] Aguilar A. L'abús de cafè ens torna més dèbils. Presència (Barcelona). Nov. 23–29, 2001; **2001**:22–23

Molecular Docking Analysis: Interaction Studies of Natural Compounds to Anti-inflammatory Targets

Rina Herowati and Gunawan Pamudji Widodo

Additional information is available at the end of the chapter

http://dx.doi.org/10.5772/intechopen.68666

Abstract

A variety of compounds from medicinal plants have been reported to possess anti-inflammatory properties. Selected natural compounds that exhibit anti-inflammatory properties were subjected to docking simulation using AutoDock Vina to investigate their interaction modes to the potential macromolecular targets. The docking was performed using different molecular targets, i.e., cyclooxygenase-2, phospholipase A2, NF-κB inhibitor, and interleukin-1 receptor. It revealed that flavonoids have the highest affinity to the macromolecular targets (the lowest binding energy values) and the highest consistency of interaction model. Some terpenoids were identified to have potential inhibitor of phospholipase A2.

Keywords: molecular docking analysis, natural compounds, anti-inflammatory, Autodock Vina

1. Introduction

Inflammation is the body defense system in response to the pathogens and injury. During the inflammation process, various inflammatory mediators are synthesized and secreted from cells and generate many cellular effects [1]. Uncontrolled inflammation lead to several chronic diseases such as cardiovascular disease, arthritis, asthma, type 2 diabetes mellitus, and cancer. To relieve inflammatory responses, nonsteroidal anti-inflammatory drugs (NSAIDs) and steroids are widely used. NSAIDs possess anti-inflammatory effect by inhibit cyclooxygenase (COX) enzyme. However, their long-term use causes gastrointestinal toxicity due to nonselective inhibition of COX-1 and COX-2 [2]. Glucocorticoids decrease the transcription of proinflammatory cytokines and chemokines and increase the transcription of anti-inflammatory cytokines, resulting in strong anti-inflammatory activity. However, their benefits are limited by the variety of systemic side effects and the development of resistance after chronic use. Thus, developing new drug candidates from natural products is greatly interesting [3].

In medicinal plants, many natural compounds, such as flavonoids, terpenoids, alkaloids, and saponin, have been reported to have in vitro as well as in vivo anti-inflammatory activity [4]. The mechanism of action and molecular target of various natural compounds needs to be studied for constructing a structure activity relationship. Molecular docking analysis can be conducted to study the interaction of these natural compounds with various molecular targets of anti-inflammatory activity. Further, the structure-activity relationship can be used to develop new derivative natural compounds with higher anti-inflammatory activity. This research aims to determine the model of interactions between the natural compounds with anti-inflammatory molecular target by molecular docking analysis. These natural compounds were used as the subjects in this study, with the cyclooxygenase-2 (COX-2), phospholipase A2, NF-κB–inducing kinase (NIK), and interleukin receptor (IRAK) as the molecular targets. The molecular docking analysis was conducted using Autodock Vina.

2. Molecular targets of anti-inflammatory agents

2.1. Cyclooxygenase-2 (COX-2)

The COX enzymes (COX-1 and COX-2) catalyze the biosynthesis of prostaglandins, prostacylins and thromboxanes, from arachidonic acid. COX-1 is constitutively expressed in most tissues, while COX-2 is expressed in specific tissues and is induced by cytokine and growth hormones. COX-1 possesses regulatory effects on platelet aggregation and gastric mucous biosynthesis, while COX-2 is involved in pathological conditions such as inflammation, pain, and fever. NSAIDs possess their anti-inflammatory activity by inhibition of COX-1 and COX-2. Prolonged inhibition of COX-1 in the gastrointestinal system causes gastrointestinal tract injury due to ulcer formation and gastric bleeding. Coxibs, the COX-2 selective inhibitor, were designed to inhibit COX-2 over COX-1, to obtain desired anti-inflammatory activity with minimal gastric toxicity side effect [5].

COX-1 and COX-2 were almost identical, despite of the residues of Ile434, His513, and Ile523 in COX-1, while in COX-2 were Val434, Arg513, and Val523. These differences result in a volume increasing of the COX-2 active site and additional side pocket off the main channel. The structures of coxibs consist of diaryl heterocycle with a sulfonamide or methyl sulfone moiety, which will bind to the side pocket of COX-2 to provide isoform-selective inhibition [6].

2.2. Phospholipase A2 (PLA2)

Phopholipase A2 (PLA2) enzymes are required to increase the level of arachidonic acid for metabolism and biosynthesis of eicosanoid under physiological condition as well as in inflammatory cell activation. PLA2 superfamily consists of cytosolic calcium dependent PLA2 (cPLA2), cytosolic calcium-independent PLA2 (iPLA2), and secreted PLA2 (sPLA2). iPLA2 is

http://dx.doi.org/10.5772/intechopen.68666

constitutively generating a low level of free fatty acids with relatively minimal specificity for the particular esterified fatty acid. cPLA2 hydrolysis arachidonic acid-containing phospholipids, leading to the production of proinflammatory eicosanoids. sPLA2 is an inducible enzyme that augments cPLA2 function to control the magnitude and duration of elevated free fatty acid levels including arachidonic acid [7].

2.3. NF-κB–inducing kinase (NIK)

Nuclear factor (NF)-κB is a group of eukaryotic transcription factors that regulates the expression of gene important for immune responses. NF-κB-inducing kinase (NIK) activates NF-κB2 by promoting proteolytic processing and the generation of NF-κB transcription of the targeted gene. NIK is also required in the signaling pathways elicited by other cytokines. NIK regulates both inflammation-induced and tumor-associated angiogenesis. NIK is highly expressed in endothelial cells of tumor tissues and inflamed rheumatoid arthritis synovial tissues [8].

2.4. Interleukin-1 receptor-associated kinase-4 (IRAK-4)

Contribution of interleukin-1 (IL-1), a proinflammatory cytokine, in the inflammation network is important. It propagates and amplifies signals; furthermore, the signaling pathways mediated by IL-1 and other cytokines receptors may communicate in various cross-talk mechanisms. Therefore, inhibition of IL-1 receptor would have profound effects on overall inflammatory responses. Interleukin-1 receptor-associated kinase 4 (IRAK-4) plays a pivotal role in signaling cascades associated with the immune and inflammatory diseases, and may be an effective therapeutic target for various diseases associated with deregulated inflammation [9].

3. Anti-inflammatory activity of natural compounds

3.1. Flavonoid and phenolic compounds

Flavonoids belong to a group of natural compounds and occur as aglycone, glycosides, and other derivatives. The flavonoids are categorized into flavonols, flavones, cathecins, flavanones, anthocyanidins, and isoflavonoids. Flavonoids exert their anti-inflammatory activity by various mechanisms, i.e., inhibition of phospholipase A2, COX, and LOX. Other mechanisms include inhibition of histamine release, phosphodiesterase, protein kinases, and activation of transcriptase. Phenylated flavonoids and a number of biflavonoids (amentoflavone, bilobetin, morelloflavone and ginkgetin) have been shown to inhibit phospholipase C1 and A2. Quercetin is reported as a strong inhibitor of both COX-2 and 5-LOX [10].

Curcumin, a polyphenol compound derived from the rhizomes of the plant turmeric, has anti-inflammatory activity, mainly due to inhibition of arachidonic acid (AA) metabolism, cyclooxygenase (COX), lipoxygenase (LOX), cytokines interleukin (IL) and tumor necrosis factor (TNF), and nuclear factor kappa B (NF-κB), despite it is also reported to stabilize lysosomal membranes [11–14].

3.2. Terpenoids

Terpenoids are classified into hemi-, mono-, sesqui-, di-, sester-, tri- and tetraterpenoids. A large numbers of terpenoids have been tested for anti-inflammatory properties. Anti-inflammatory activity of 1,8-Cineol, a monoterpene oxide, is correlated to inhibition of leukotriene B4, prostaglandin E_2 TNF-α, interleukin, and thromboxane [15]. Parthenolide, a sesquiterpene lactone, possesses anti-inflammatory activity by several mechanisms, including inhibition of NF-κB [16]. Cucurbitacins, a group of triterpenes, were reported to show anti-inflammatory activity by inhibition of prostaglandin production and blocking NF-κB activation [17].

3.3. Alkaloid

Alkaloids are the basic substances that contain one or more nitrogen atoms, usually in combination as part of a cyclic system. They are often toxic to have various pharmacological activities, including anti-inflammatory activity [18]. Isoquinoline, quinoline, and indole alkaloids were the most studied classes for anti-inflammatory activity. Berberine, an isoquinoline alkaloid, showed potential in vitro and in vivo anti-inflammatory activity. It was reported to inhibit prostaglandin E2 production, without inhibitory effect on either COX-1 or COX-2 activity [19].

3.4. Saponins

Saponins are group of glycosides found in many plants, with triterpenoid or steroid as the aglycone moiety. Kalopanaxsaponin A was triterpenoid saponin isolated from the stem bark of *Kalopanax pictus* that showed significant anti-inflammatory activity [20]. Loniceroside A, a triterpenoid saponin isolated from the aerial parts of *Lonicera japonica*, also showed comparable anti-inflammatory activity to aspirin. It exhibited anti-inflammatory activity against acute and chronic inflammation [21].

4. Molecular docking analysis

Molecular modeling investigations were carried out using Zyrex Cruiser workstation EM4100 running Intel Core i3-4030U Processor, 2 GB RAM, 500 GB hard disk, and Intel HD Graphic Family graphics card. Autodock Vina docking program, Molecular Graphic Lab, The Scripps Research Institute [22] was employed for the docking studies.

4.1. Preparation of target proteins

PDB structures used, i.e., 5IKT (COX-2), 4UY1 (PLA2), 1DV4 (NF-κB), 5KX7 (IRAK-4), were obtained from the Brookhaven Protein Data Bank (www.rcsb.org). **Table 1** presented the

http://dx.doi.org/10.5772/intechopen.68666

Target macromolecule (pdb code)	Ligand	Structure
COX-2 (5IKT)	Tolfenamic acid	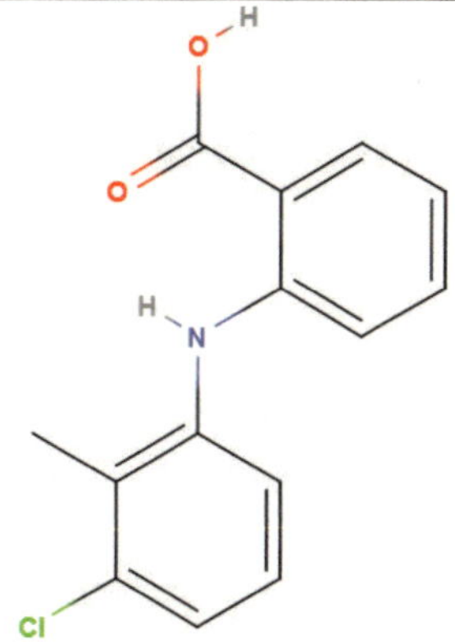
Phospholipase A2 (4UY1)	5-(2,5-dimethyl-3-thienyl)-1H-pyrazole-3-carboxamide	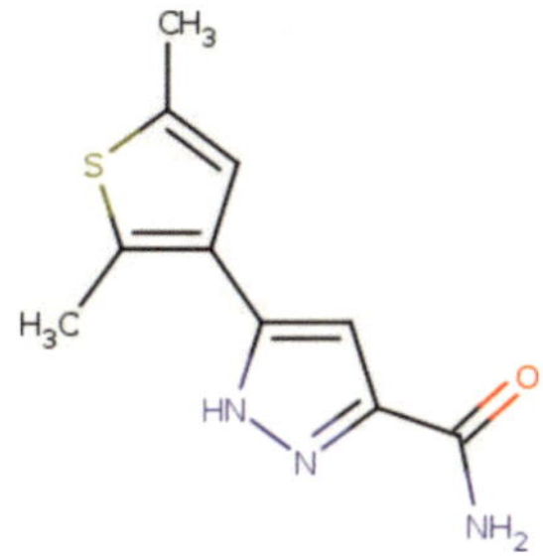
NIK (IDV4)	4-{3-[2-amino-5-(2-methoxyethoxy)pyrimidin-4-yl]-1H-indol-5-yl}-2-methylbut-3-yn-2-ol	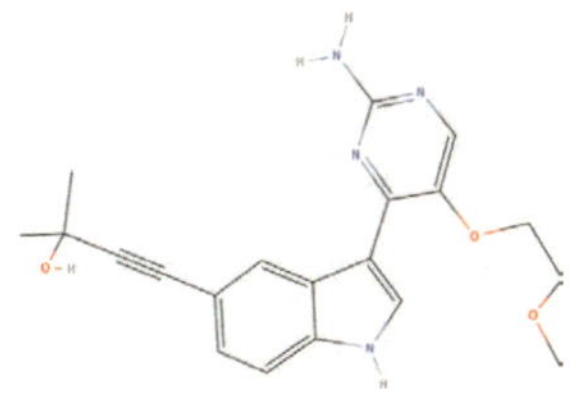
IRAK-4 (5KX7)	~{N}-(3-aminocarbonyl-1-methyl-pyrazol-4-yl)-6-(1-methylpyrazol-4-yl) pyridine-2-carboxamide	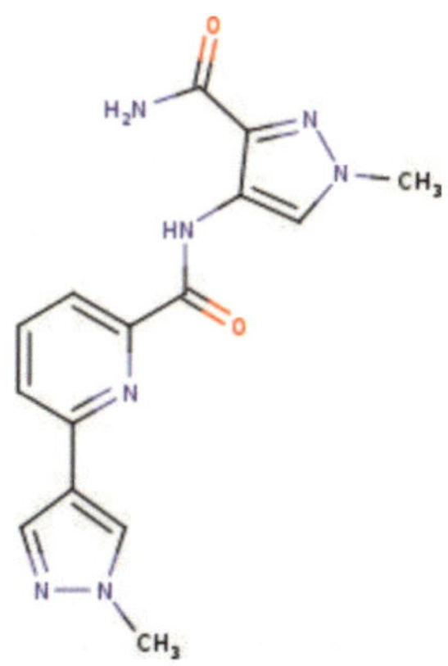

Table 1. Target macromolecules, pdb codes, native ligands, and ligand structures used in the docking study.

target macromolecules and each of the native ligand. The protein structures were prepared using UCSF Chimera 1.11.2 to remove all nonreceptor atom including water, ion, and miscellaneous compounds. The obtained structures then were saved as pdb file.

4.2. Preparation of ligands

The structures of native ligands from each target macromolecules were prepared by UCSF Chimera 1.11.2 to separate from the protein, water, and miscellaneous substances. The structures of the 40 ligands of natural compounds were sketched using MolView 2.2. Each structure then was executed an MMFF94 energy minimization. These obtained conformations were used as starting conformations to perform docking analysis.

4.3. Docking method validation

To ensure that the docking studies were valid and represented the reasonable potential binding model, the docking methods and parameters used were validated by redocking experiment. Each copy of native ligand was docking into the native protein to determine the ability of Autodock program to reproduce the orientation and position of the ligand observed in the crystal structure. The valid criteria used is the all atom root mean square deviation (RMSD) between the docked position and the crystallographically observed binding position of the ligand, and success is typically regarded as being less than 2 Å.

4.4. Docking studies

Docking studies were carried out using the above mentioned prepared target macromolecules and natural compound ligands (1–40) by employing Autodock Vina program. Docking was performed to obtain a population of possible conformations and orientations for the ligand at the binding site. The protein was loaded in PyRx software, creating a PDBQT file that contains a protein structure with hydrogens in all polar residues. All bonds of ligands were set as rotatable. All calculations for protein-fixed ligand-flexible docking were done using the Lamarckian Genetic algorithm (LGA) method. The docking site on protein target was defined by establishing a grid box with a default grid spacing, centered on the position of native ligand. The best conformation was chosen with the lowest binding energy, after the docking search completed. The interactions complex protein-ligand conformations, including hydrogen bonds and the bond lengths were analyzed using Discovery Studio Visualizer 16.1.0.15350.

5. Result and discussion

The molecular docking analysis was performed using Autodock Vina program with 40 plant-derived compounds, including flavonoids and phenolic compounds, alkaloid, terpenoid, and saponin. First, the validation method was conducted to ensure the capability of docking machine. **Figure 1** represented the validation result of docking protocol. All of the redocking ligands showed similar conformation with the native redocking ligands, and the RMSD values were ≤2.0 Å.

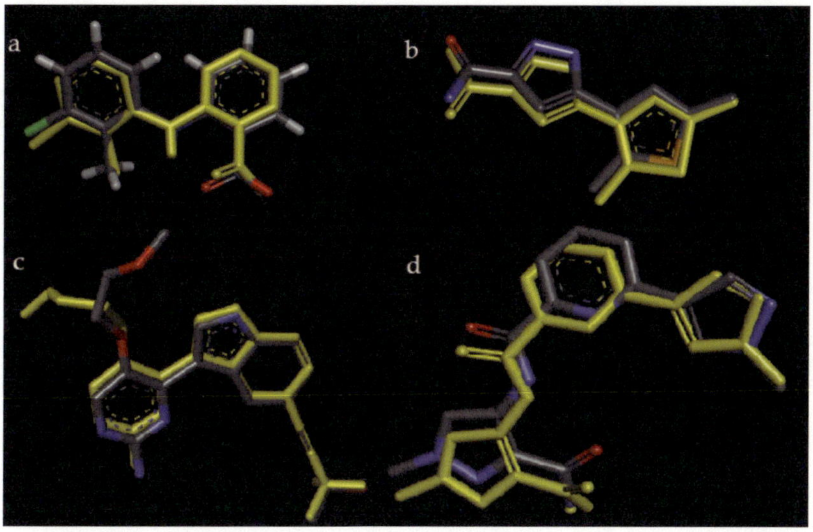

Figure 1. Overlay of the native ligands (gray) and redocking conformations (yellow). (a: COX-2; b: PLA2; c: NF-κB; d: IRAK-4).

Each natural compound then was docked into each of six different targets, using valid parameter method. The lowest energy docked conformation of the best cluster was selected and analyzed. **Table 2** summarizes the docking study results, presented as binding energy.

5.1. Interaction to Cyclooxygenase-2

Structure of COX-2 complex with tolfenamic acid, a selective COX-2 inhibitor, was used for this study. The carboxylic group of tolfenamic acid interacted by hydrogen bonding with Tyr-385A and Ser-530A at the top of channel. While the methylphenyl aminobenzoic moiety interacted hydrophobically with Val116A, Val349A, Leu352A, Val523A, Ala527A, as well as Leu531A [23]. This study revealed that some flavonoids and phenolic compounds, i.e., amentoflavone, apigenin, bilobetin, diosmine, epicatechin gallate, ginkgetin, hesperidin, luteolin, morelloflavon, and quercetin, showed lower binding energy than that of tolfenamic acid, the selective COX-2 inhibitor.

The model interaction of these flavonoids to the binding site of COX-2 was similar to the model interaction of native ligand. Carbonyl group at C-ring of flavonoid played an important role in the ligand-target interaction, by hydrogen bond interaction to Ser530A and Arg120A residue, while A- and B-phenolic ring interacted to Val349A, Leu352, Ala527, and Leu531A residue via hydrophobic interaction (**Figure 2**).

This result was in line with previous reports about COX-2 inhibitory activity of flavonoid and phenolic compounds. Several research reports about COX inhibitory activity of flavonoids were documented [24–26]. A structure-activity relationship study about COX-2 inhibitory activity of flavonoid resumed the pharmacophores group were 4-oxo (C-ring), 7-hydroxyl moiety (A-ring), as well as para-substituted phenolic B-ring [27]. Apigenin meets all the requirements, due to its hydroxylation pattern at 5′-, 7′-, and 4′-positions.

Compound	Target macromolecules; binding energy (kcal/mole)			
	COX-2	PLA2	NIK	IRAK-4
Native ligand	−9.0	−7.8	−9.9	−9.0
Flavonoids and phenolic compounds:				
Amentoflavone	−8.7	−9.0	−8.2	−8.1
Anthocyanin	−8.3	−7.8	−8.3	−8.1
Apigenin	−9.2	−7.8	−9.1	−9.3
Apocynin	−6.7	−6.5	−6.8	−6.3
Apocynin ester	−6.2	−5.8	−6.0	−6.2
Bilobetin	−10.6	−9.1	−6.3	−10.1
Curcumin	−8.9	−8.4	−9.0	−9.3
Diosmine	−8.6	−8.9	−5.6	−8.8
Epicatechin gallate	−9.9	8.2	−7.9	−8.4
Epigallocatechin gallate	−8.4	8.1	−8.0	−8.8
Ginkgetin	−9.4	−9.0	−6.4	−10.2
Hesperidin	−8.8	−8.9	−6.1	−8.2
Hydrocinnamic acid	−6.4	−6.6	−6.4	−6.8
Luteonin	−9.4	−8.1	−9.3	−8.9
Morelloflavon	−9.5	−8.1	−4.1	3.0
Paeonol	−6.5	−6.0	−5.9	−6.1
Procyanidin	−7.9	−8.5	−6.5	−4.9
Pterostilben	−7.6	−6.7	−8.0	−8.0
Quercetine	−9.5	−8.1	−8.9	−9.1
Resveratol	−7.9	−7.1	−8.0	−8.1
Yuccaol A	−7.8	−8.2	−6.4	0.3
Yuccaol B	−8.0	−7.0	0.3	−0.8
Yuccaol C	−7.8	−6.9	−0.4	−0.5
Yuccaol D	−8.5	−7.5	−1.8	3.3
Yuccaol E	−8.1	−7.1	−0.9	−1.0
Alkaloids				
Berberine	−9.8	−8.4	−8.9	−9.3
Terpenoids				
1.8-Cineol	−5.6	−5.2	−5.8	−5.9
Cucurbitacin B	−8.0	−7.6	−2.1	−0.6

http://dx.doi.org/10.5772/intechopen.68666

Compound	Target macromolecules; binding energy (kcal/mole)			
	COX-2	PLA2	NIK	IRAK-4
Cucurbitacin D	−8.2	−7.4	−3.3	−1.7
Cucurbitacin E	−8.8	−7.8	−2.9	−3.3
Cucurbitacin I	−8.4	−7.3	−3.7	−3.0
Cucurbitacin R	−8.3	−7.9	−2.6	−1.3
Dihydrocucurbitacin B	−7.5	−7.2	−2.1	−1.8
Oleanolic acid	−8.7	−7.5	1.3	−3.0
Partenolid	−7.7	−8.1	−6.7	−7.3
Pseudopterosin	−7.8	−7.5	−8.3	−9.0
Ursolic acid	−7.8	−8.0	1.7	−3.6
Saponins				
Kalopanaxsaponin	−8.7	−7.7	9.1	−0.5
Loniseroside A	−6.1	−7.8	−7.4	−6.5

Table 2. Binding energy values of natural compounds docked to target macromolecules.

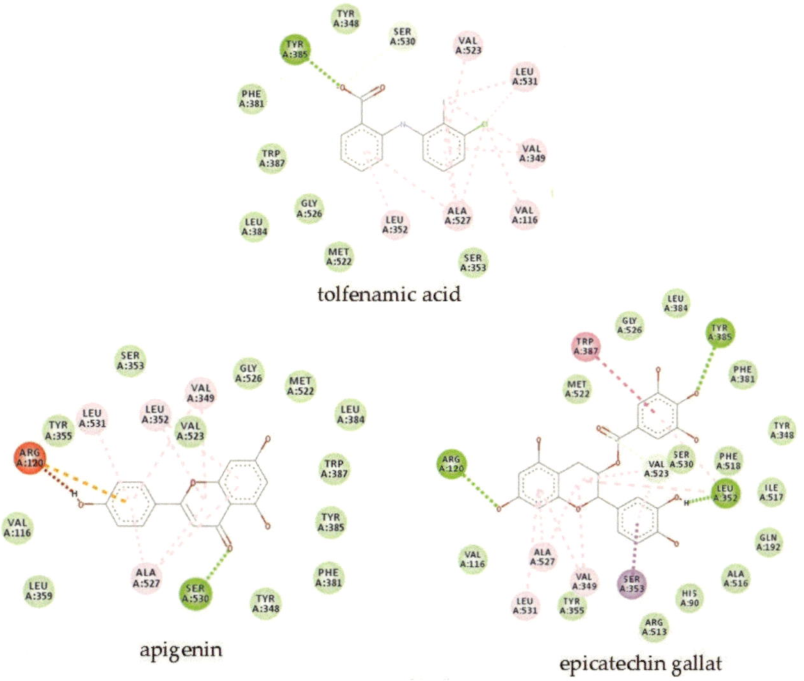

Figure 2. Model interaction of tolfenamic acid, apigenin and epicatechin gallate with COX-2 binding site. Dark green: hydrogen bond; light green: van der Waals; red: electron donor-donor interaction; pink-violet: hydrophobic interaction.

Epicatechin gallate was reported to inhibit COX-2 enzyme [28]. In this study, epicatechin gallate also showed good interaction to COX-2 binding site, where hydrogen bond was formed between its phenolic groups and Tyr385A as well as Arg180A residues, in addition to hydrophobic interactions between the aromatic groups and Val349A, Val523A, Ala527A, and Leu531A. Berberine, the alkaloid used in study, showed lower affinity compared to native ligand. This result was in agreement with the fact that berberine does not interfere with COX enzyme activity [19].

5.2. Interaction to Phospholipase A2

Complex structure of PLA with 5-(2,5-dimethyl-3-thienyl)-1H-pyrazole-3-carboxamide, shows that the terminal amide group interacts with by hydrogen bonding with Asp47A and His46A, while hydrophobic interaction is formed between the thienyl pyrazole moiety and Leu5A [26]. Three biflavonoids, i.e., amentoflavone, bilobetin, and ginkgetin, were the ligands with the lowest binding energy, with the binding energy value being −9.0, −9.1, and −9.0 kcal/mole respectively. Biflavanoids were repeatedly reported as PLA2 inhibitors [29–32].

The biflavonoids successfully docked into the PLA2 binding site (**Figure 3**). The compounds bind to the target via hydrogen bond with Leu29A residue, while hydrophobic interactions were formed between aromatic rings of biflavonoid with Ala2A, Leu5A as well as Ala6A. Hydrophobic pharmacophores of biflavonoids support the PLA2 inhibitory activity, due to interaction between substrate of PLA2 usually occurs at hydrophobic channel of the enzyme.

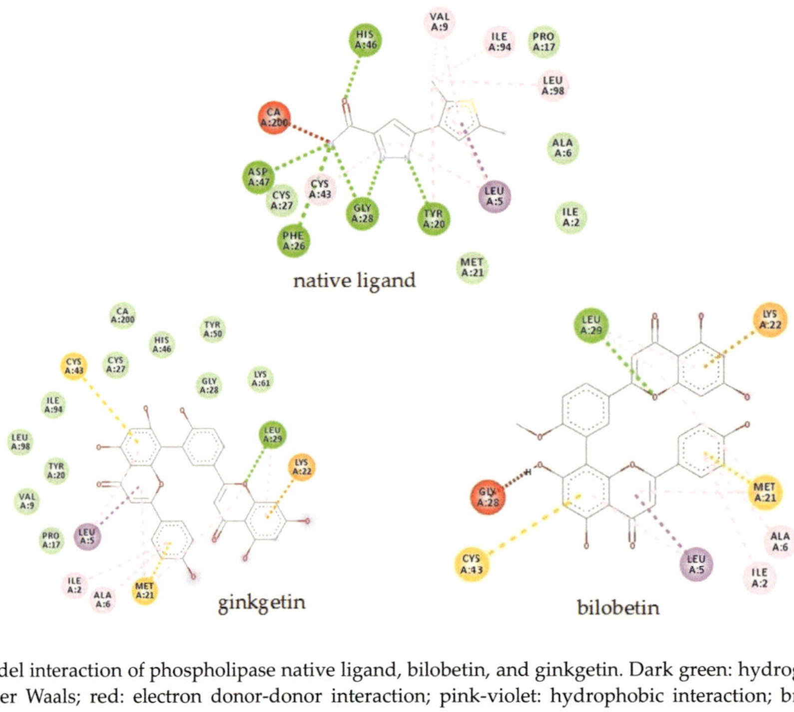

Figure 3. Model interaction of phospholipase native ligand, bilobetin, and ginkgetin. Dark green: hydrogen bond; light green: van der Waals; red: electron donor-donor interaction; pink-violet: hydrophobic interaction; brown: π-cation interaction.

The π-cation interaction increases the affinity of biflavonoid, because the Ca^{2+} ion is important for the catalytic mechanism of PLA2 [33].

5.3. Interaction to NF-κB–inducing kinase (NIK)

Study about NIK inhibitory has been conducted with the analogs of imidazopyridinyl pyrimidinamine. The hydrogen bonds were formed between pyrimidinamine group and Glu470A as well as Leu 472A; furthermore, imidazopyridinyl pyrimidinamine ring was hydrophobically interacted to Val414A, Ala427A, Lys429A, Leu522A, and Cys533A. Interaction of π-sulfur was formed between pyrimidinamine ring and Met469A [34].

This study revealed that apigenin and luteolin, two flavonoid aglikons, had lowest binding energy values. This was in line with previous research reports that apigenin and luteolin inhibit activation of NF-πB [35–38].

The similarity of interaction model between native ligand and these flavonoids are presented in **Figure 4**. Hydrogen bond was formed between carbonyl group at C-ring (4-oxo) of flavonoid and Ser476A residue. Similar to the interaction model of native ligand, hydrophobic interactions were formed between aromatic rings of flavonoid with Val414A, Ala427A, and Leu522A residues. Interaction of π-sulfur was also formed between B-ring with Met469A residue. Additional hydrogen bonding was formed between hydroxyl groups of B-ring with Glu470A and Leu472A residues. This result concluded that *o*-catechol or *p*-hydroxy B-ring

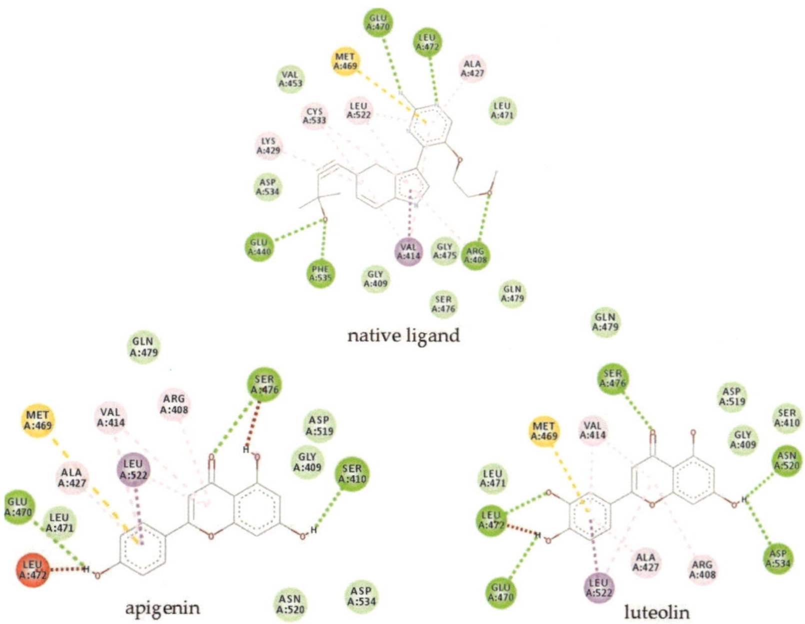

Figure 4. Model interaction of NIK native ligand, apigenin, and luteolin. Dark green: hydrogen bond; light green: van der Waals; red: electron donor-donor interaction; pink-violet: hydrophobic interaction; brown: π-sulfur interaction.

was the pharmacophore group for NIK inhibitory activity, in addition to -oxo and planar aromatic ring of flavonoids.

5.4. Interaction to Interleukin-1 receptor-associated kinase-4 (IRAK-4)

The X-ray crystal structure of IRAK-4-inhibitor complex has been reported. The structure revealed that the carboxamide group and N-pyrazol contributed as hydrogen bond acceptor (interacted to Met265B and Lys213B, respectively). Furthermore, two aromatic rings acted as hydrogen bond donor increased the affinity to the target via hydrophobic interaction with Val200B, Ala211B, Val246B, Tyr262B, as well as Leu318B [39].

Among 40 natural compounds docked to IRAK-4, apigenin, bilobetin, curcumin, ginkgetin, quercetin, berberine, and pseudopterosin showed higher affinity compared to the native ligand. Quercetin was reported to show IRAK-4 inhibitory activity [40]. Affinity of quercetin to the binding site was supported by hydrogen bond between 4-hydroxy of B-ring with Lys213B and Glu233B residues, while the aromatic rings contributed by forming hydrophobic interaction with Val200B, Tyr262B, and Leu318B. Similar to the interaction model of native ligand to IRAK-4, aromatic rings of bilobetin also formed the hydrophobic interaction to Val200B, Tyr262B, and Leu318B residues of IRAK-4 (**Figure 5**).

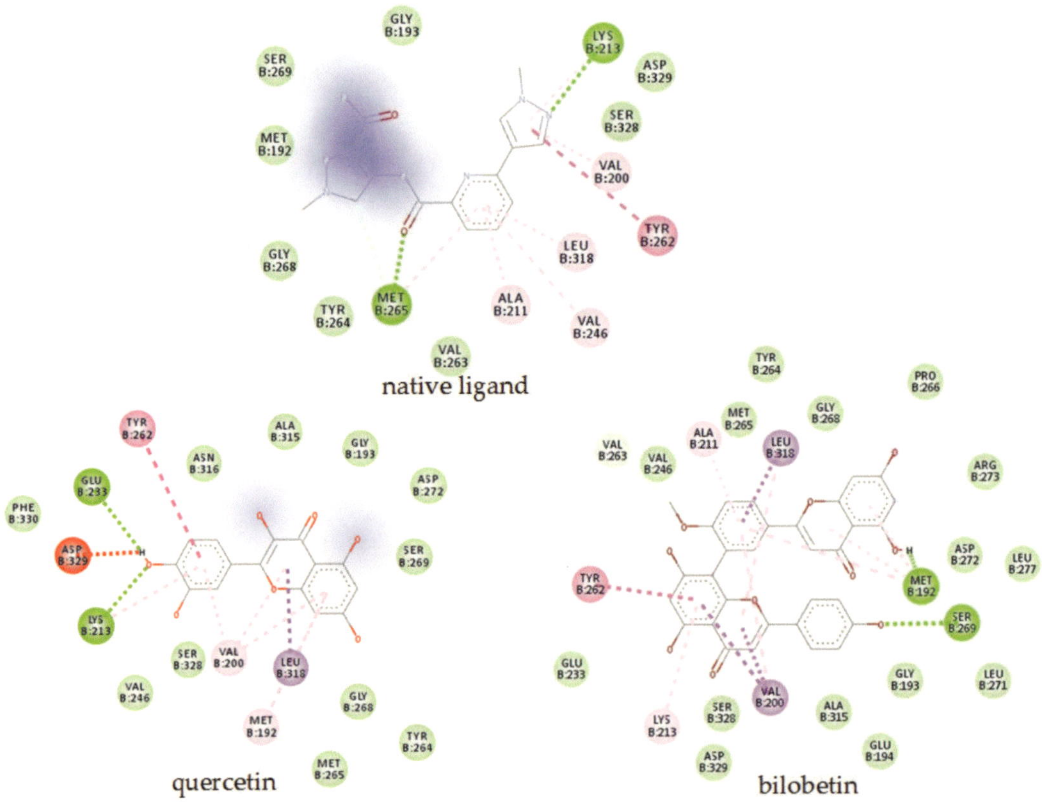

Figure 5. Model interaction of NIK native ligand, quercetin, and bilobetin. Dark green: hydrogen bond; light green: van der Waals; red: electron donor-donor interaction; pink-violet: hydrophobic interaction.

6. Conclusion

The docking results revealed that the highest affinity to the macromolecular targets (the lowest binding energy values) and the consistency of interaction model were shown by the flavonoids. Some terpenoids were identified to have potential inhibitor of phospholipase A2.

Acknowledgements

This work was funded by the Ministry of Research Technology and Higher Education, Republic of Indonesia.

Author details

Rina Herowati[1*] and Gunawan Pamudji Widodo[2]

*Address all correspondence to: rn_herowati@yahoo.co.id

1 Departement of Pharmacochemistry, Faculty of Pharmacy, Setia Budi University, Indonesia

2 Departement of Pharmacology, Faculty of Pharmacy, Setia Budi University, Indonesia

References

[1] Azab A, Nassar A, Azab AN. Anti-inflammatory activity of natural products. Molecules 2016;(21):1321. DOI: 10.3390/molecules21101321

[2] Sostres C, Gargallo CJ, Arroyo MT, Lanas A. Adverse effects of non-steroidal anti-inflammatory drugs (NSAIDs, aspirin and coxibs) on upper gastrointestinal tract. Best Practice & Research. Clinical Gastroenterology 2010;24:121-132. DOI: 10.1016/j.bpg.2009.11.005

[3] Allijn IE, Vaessen SFC, van Ufford LCQ, Beukelman KJ, de Winther MPJ, Storm G, et al. Head-to-head comparison of anti-inflammatory performance of known natural products in vitro. PLOS ONE. 2016;**11**(5). DOI: 10.1371/journal.pone.0155325

[4] Yuan G, Wahlqvist ML, He G, Yang M, Li D. Natural products and anti-inflammatory activity. Asia Pacific Journal of Clinical Nutrition. 2006;15(2):143-152

[5] Orlando BJ, Malkowski MG. Crystal structure of rofecoxib bound to human cyclooxygenase-2. Acta Cryst. 2016;(F72):772-776. DOI: 10.1107/S2053230X16014230

[6] Orlando BJ, Malkowski MG. Substrate-selective inhibition of cyclooxygenase-2 by fenamic acid derivatives is dependent on peroxide tone. Journal of Biological Chemistry. 2016;**291**(29):15069-15081. DOI: 10.1074/jbc.M116.725713

[7] Dennis EA, Norris P. Eicosanoid storm in infection and inflammation. Nature Reviews Immunology. 2015. DOI: 10.1038/nri385

[8] Li K, McGee LR, Fisher B, Sudom A, Liu J, Rubenstein SM, et al. Inhibiting NF-κB-inducing kinase (NIK): Discovery, structure-based design, synthesis, structure–activity relationship, and co-crystal structures. Bioorganic & Medicinal Chemistry Letters 2013;(23):1238-1244. DOI: 10.1016/j.bmcl.2013.01.012

[9] Wang Z, Wesche H, Stevens T, Walker N, Yeh C. IRAK-4 inhibitor for inflammation. Current Topics in Medicinal Chemistry 2009;**9**:724-737

[10] Rathee P, Chaudhary H, Rathee S, Rathee D, Kumar V, Kohli K. Mechanism of action of flavonoids as anti-inflammatory agents: A review. Inflammation & Allergy Drug Targets. 2009;8(3):229-235

[11] Jian YT, Wang JD, Mai GF, Zhang YL, Lai ZS. Curcumin regulates cyclooxygenase-2 activity in trinitrobenzene sulfonic acid-induced colitis. Journal of the Fourth Miltary Medical University2005;**26**:521-524

[12] Skrzypczak-Jankun E, McCabe NP, Selman SH, Jankun J. Curcumin inhibits lipoxygenase by binding to its central cavity: Theoretical and X-ray evidence. International Journal of Molecular Medicine2000;**6**:521-526

[13] Kang BY, Chung SW, Chung W, Im S, Hwang SY, Kim TS. Inhibition of interleukin-12 production in lipopolysaccharide-activated macrophages by curcumin. European Journal of Pharmacology. 1999;**384**:191-195

[14] Bremner P, Heinrich M. Natural products and their role as inhibitors of the pro-inflammatory transcription factor NF-kB. Phytochemistry Reviews. 2005;**4**:27-37

[15] Juergens UR, Stober M, Vetter H, Juergens UR, Stober M, Vetter H. Inhibition of cytokine production and arachidonic acid metabolism by eucalyptol (1,8-cineole) in human blood monocytes in vitro. European Journal of Medical Research. 1998;**3**:508-510

[16] Hehner SP, Hofmann TG, Droge W, Schmitz ML. The antiinflammatory sesquiterpene lactone parthenolide inhibits NF kappa B by targeting the Ikappa B kinase complex. Journal of Immunology. 1999;**163**:5617-5623

[17] Jian CC, Ming HC, Rui LN, Cordell GA, Qiu SX. Cucurbitacins and cucurbitane glycosides: Structures and biological activities. Natural Product Reports 2005;**22**:386-399

[18] Souto AL, Tavares JF, da Silva MS, Diniz MFFM, Athayde-Filho PF, Filho JMB. Anti-inflammatory activity of alkaloids: An update from 2000 to 2010. Molecules. 2011;**16**:8515-8534. DOI: 10.3390/molecules16108515

[19] Kuo C, Chi C, Liu T. The anti-inflammatory potential of in vitro and in vivo. Cancer Letters. 2004;**203**:127-137. DOI: 10.1016/j.canlet.2003.09.002

[20] Choi JW, Huh K, Kim SH, Lee KT, Lee HK, Park HJ. Kalopanaxsaponin A from Kalopanax pictus, a potent antioxidant in the rheumatoidal rat treated with Freund's complete adjuvant reagent. Journal of Ethnopharmacology. 2002;79:113-118

[21] Lee SJ, Son KH, Chan HW, Kang SS, Kim HP. Antiinflammatory activity of the major constituents of Lonicera japonica. Archives of Pharmacal Research. 1995;18:133-135

[22] Trott OA, Olson J. Autodock Vina: Improving the speed and accuracy of docking with a new scoring function, efficient optimization and multithreading. Journal of Computational Chemistry. 2010;**31**:455-461

[23] Orlando BJ, Malkowski MG. Substrate selective inhibition of cyclooxygeanse-2 by fenamic acid derivatives is dependent on peroxide tone. Journal of Biological Chemistry. 2016;**291**:15069-15081. DOI: 10.1074/jbc.M116.725713

[24] Kim HP, Son KH, Chang HW, Kang SS. Anti-inflammatory plant flavonoids and cellular action mechanisms. Journal of Pharmacological Sciences. 2004:229-249

[25] Kartasasmita RE, Herowati R, Harmastuti N, Gusdinar T. Quercetin derivatives docking based on study of flavonoids interaction to cyclooxygenase-2. Indonesian Journal of Chemistry. 2009;**2**:297-302

[26] Riberio D, Freitas M, Tome SM, Silva AM, Laufer S, Lima JL, Fernanes E.Flavonoids inhibit COX-1 and COX-2 enzymes and cytokine/chemokine production in human whole blood. Inflammation. 2015;**38**(2):858-870. DOI: 10.1007/s10753-014-9995-x

[27] Rosenkanz HS, Thampatty BP. SAR: Flavonoids and COX-2 inhibition. Oncology Research. 2003;**12**:529-535

[28] Seeram NP, Zhang Y, Nair MG. Inhibition of proliferation of human cancer cells and cyclooxygenase enzymes by anthocyanidins and catechins. Nutrition and Cancer. 2003;**46**(1):101-106. DOI: 10.1207/S15327914NC4601_13

[29] Chen H, Knerr L, Åkerud T, Hallberg K, Öster L, Rohman M, Beisel HG, Olsson T, Brengdhal J, Sandmark J, Österlund K, Bodin C. Discovery of novel pyrazole series of group X secreted phospholipase A2 inhibitor (sPLA2X) via fragment based virtual screening. Bioorganic & Medicinal Chemistry 2004. DOI: 10.1016/j.bmcl.2014.09.058

[30] Lee SJ, Son KH, Chang HW, Kang SS, Kim HP. Inhibition of arachidonate release from rat peritoneal macrophages by biflavonoid. Archives of Pharmacal Research. 1997;**20**:533-538

[31] Baek SH, Yun SS, Kwon TK, Kim JR, Chang HW, Kwak JY, Kim JH, Kwun KB. The effects of two new antagonists of secretory PLA 2 on TNF-, iNOS, and COX-2 expression in activated macrophages. Shock. 1999;**12**:473-478

[32] Kim HP, Pham HT, Ziboh VA. Flavonoids differentially inhibit guinea pig epidermal cytosolic phospholipase A2. Prostaglandins, Leukotrienes, and Essential Fatty Acids. 2001;**65**:281-286

[33] Dileep KV, Tintu I, Sadasivan C. Molecular docking studies of curcumin analogs with phospholipase A2. Interdisciplinary Sciences: Computational Life Sciences. 2011;**3**:189-197. DOI: 10.1007/s12539-011-0090-9

[34] Hanisak J, Michael Seganish W, McElroy WT, Tang H, Zhang R, Tsui H-C, Fischmann T, Tulshian D, Tata J, Sondey C, Devito K, Fossetta J, Garlisi CG, Lundell D, Niu X Efforts

towards the optimization of a bi-aryl class of potent IRAK4 inhibitors, Bioorganic & Medicinal Chemistry Letters. 2016. DOI: 10.1016/j.bmcl.2016.07.048

[35] Nam NH. Naturally occuring NF-kappaB inhibitors. Mini Reviews in Medicinal Chemistry. 2006;**6**(8):945-951

[36] Ruiz PA, Haller D. Functional diversity of flavonoids in the inhibition of the proinflammatory NF-kB, IRF, and Akt signaling pathways in murine intestinal epithelial cells. The Journal of Nutrition. 2005:664-671

[37] Wang J, Liu Y, Xiao L, Zhu L, Wang Q, Yan T. Anti-inflammatory effects of Apigenin in lipopolysaccharide-induced inflammatory in acute lung injury by suppressing COX-2 and NF-kB pathway. Inflammation. 2014;**37**(6):2085-2090. DOI: 10.1007/s10753-014-9942-x

[38] Chen C, Peng W, Tsai K, Hsu S. Luteolin suppresses inflammation-associated gene expression by blocking NF-κB and AP-1 activation pathway in mouse alveolar macrophages. Life Sciences. 2007;**81**:1602-1614. DOI: 10.1016/j.lfs.2007.09.028

[39] Zhong L, Zhou L, Tian Y, You R. Identification of novel IRAK-4 inhibitors through pharmacophore modeling, atom based 3D-QSAR docking strategies and molecular dynamic simulation. Letters in Drug Design and Discovery. 2016;**13**(9):879-887. DOI: 10.2174/1570180813666160421163027

[40] Shibata T, Nakashima F, Honda K, Lu Y, Kondo T, Ushida Y, et al. Toll-like receptors as a target of food-derived anti-inflammatory compounds. The Journal of Biological Chemistry. 2014;**289**(47):32757-32772. DOI: 10.1074/jbc.M114.585901